DESIGN OF LARGE STEAM TURBINE-GENERATOR FOUNDATIONS

Prepared by the
Task Committee on Turbine Foundations
of the Fossil Power Committee and the
Nuclear Energy Committee of the
Energy Division of the
American Society of Civil Engineers

February 1987

ASCE 1852

Published by the
American Society of Civil Engineers
345 East 47th Street
New York, New York 10017-2398

ABSTRACT

This report is a guide for engineers who are responsible for the design and analysis of large turbine-generator foundations. A description of the various components of turbine-generator equipment is presented. Load cases that must be considered for analysis and design of the foundation are discussed. Guidelines are provided for establishing initial overall foundation dimensions, and for establishing the initial sizes of the transverse and longitudinal girders, columns, and foundation mat. Development of both static and dynamic analytical models for the turbine foundation, including the mat and supporting soil, is discussed. A commentary is given on the various deflection criteria utilized and the load cases to be considered for evaluating deflections. Design criteria for the components of the foundation are reviewed. A general discussion of steel foundations is presented. The concept of spring-mounted machine foundations and their details are reviewed.

Library of Congress Cataloging-in-Publication Data

Design of large steam turbine-generator foundations.

"February 1987."
Bibliography: p.
1. Turbogenerators—Foundations—Design and construction. I. American Society of Civil Engineers. Task Committee on Turbine Foundations.
TK2458.F68D47 1987 621.1'97 87-1243
ISBN 0-87262-597-4

Library of Congress Catalog Card No.: 87-1243
ISBN 0-87262-597-4
Manufactured in the United States of America.

Design of Large Steam Turbine-Generator Foundations

Task Committee on Turbine Foundations

Fossil Power Committee
Nuclear Energy Committee
Energy Division

R. J. Small, Chairman

Ihsan U. Bari
Peter C. Charoglu
Ashak J. Desai
James Ferrara
Ray R. Funk
James L. Goy
Asadour Hadjian
Wolf Jeske
John S. Joyce
Robert McDonald
Joseph Mora
Norman Morantz
Anthony Niemzak
S. N. Pagay
Garry J. Patterson
A. K. Singh
Joseph T. Sollecito
Russell L. Soper
Garland W. Steinmann
Robert E. Stippich
Edward M. Suplick
Francis M. Thomas
Donald Walters
Matthew W. Wrona

CONTENTS

Chapter 1

Introduction

The purpose of the Task Committee was to prepare a guide for the analysis and design of large turbine-generator foundations.

The membership of the Committee consisted of representatives of 13 engineering or engineering-constructor firms involved either solely or principally in the design of power plants, 4 turbine-generator manufacturers, 3 utilities, and 1 university. This membership from the various segments of the power industry provided a good cross section and depth of engineering knowledge to address the variety of problems confronting the foundation designer.

Since, with very few exceptions, foundations in the United States are constructed of concrete, it was initially decided to limit this report to the consideration of concrete foundations. However, because the members of the power industry have demonstrated a continued interest in steel foundations, a general discussion of this type of support is included in the report. Since spring-supported turbine-generator sets are becoming more common in several European and other countries, a brief discussion of this type of support is also presented.

This report attempts to organize into a single document all the criteria necessary to analyze and design a large turbine-generator foundation. In addition to the more common concepts utilized in the design process, the report has attempted to address several areas that have, in the past, caused designers some difficulty. These include differential machine deflections, differential thermal expansion of the foundation, foundation creep and shrinkage, and soil-foundation interaction. Other areas addressed are machine-foundation-soil dynamics, mathematical and analytical modeling procedures, and design and detailing procedures.

Specifically, the report is organized into nine chapters. The order of the chapters generally follows the same order that the engineer would follow when compiling information for the design of the foundation. A brief summary of the main subject of each chapter is provided below.

- Chapter 2, Turbine-Generator Equipment, presents a brief description of the various components of the machine to provide a better understanding of how the machine functions, thus providing a greater appreciation of the manufacturers' design criteria.

- Chapter 3, Loads, discusses all the load cases that must be considered for the analysis and design of the foundation.

- Chapter 4, Foundation Geometry, provides guidelines for establishing initial overall foundation dimensions, and for establishing the initial sizes of the transverse and longitudinal girders, columns, and foundation mat.

- Chapter 5, Analysis Techniques, provides a discussion of the development of both static and dynamic analytical models for the turbine foundation, including the mat and supporting soil.

- Chapter 6, Deflections, presents a commentary on the various deflection criteria utilized and the load cases to be considered for evaluating deflections.

- Chapter 7, Design, discusses the design criteria for the different components of the foundation, including code interpretations where required.

- Chapter 8, Steel Foundations, presents a general discussion of steel foundations.

- Chapter 9, Spring-Mounted Machine Foundations, discusses the concept of spring foundations and presents typical details associated with this type of support.

This report does not attempt to explain or clarify specific manufacturers' design criteria. This was decided because these criteria change frequently, which would quickly render parts of the report obsolete. Also, a large part of the report would have to be devoted to this topic, and it was thought better to devote the Committee's time to other subjects.

The Committee realized that in different parts of the world, and even among different firms in the United States, the various parts of the foundation are referred to by many different names. For the purpose of this report, the Committee decided to adopt the following nomenclature for the parts of the foundation.

- Longitudinal beams - those members that support the turbine-generator set that are parallel to the axis of the machine.

- Transverse beams - those members that support the turbine-generator set that are transverse to the axis of the machine.

- Deck - the uppermost portion of the foundation, consisting of the longitudinal and transverse beams.

- Mat - the part of the foundation supported on soil, rock, or piles.

- Columns - the vertical members that support the deck and are in turn supported by the mat.

- Superstructure - the part of the foundation above the mat, consisting of the columns and the deck.

- Foundation - the entire structure, including the deck, columns, and mat.

In the past, there has been much interest in what has been generally referred to as the "low-tuned" turbine foundation. This type of foundation has a fundamental vertical frequency much lower than the running speed of the machine and is characterized by extremely slender columns. However, all turbine foundations, even the relatively massive foundations constructed many years ago, were probably low-tuned. The foundations that have been identified as low-tuned are simply lower tuned than those typically constructed in the United States. For the sake of distinguishing between these two types of foundations, the Committee adopted the following definitions:

Low-tuned foundation: $f_m/f_n > 3.0$ (1-1)

Conventional foundation: $1.4 < f_m/f_n < 3.0$ (1-2)

where:

f_m = running speed of machine (cycles/second), and

f_n = fundamental vertical frequency of the foundation, assuming its base fixed (cycles/second).

When the ratio f_m/f_n is less than 0.7, the foundation is referred to as high tuned. Because high-tuned foundations are extremely massive and stiff, they are uneconomical and are seldom used.

It is appropriate to make several general comments about the low-tuned foundation concept.

It is generally not feasible to achieve as low a tuning with U.S. foundations as can be achieved in other countries. It is typical American practice to support the condenser rigidly at its base and provide an expansion joint at the turbine-condenser interface. This arrangement causes the deck to be subjected to the full condenser vacuum load. Since this load is applied after the shaft has been aligned, the foundation columns must be relatively large to satisfy the manufacturer's deflection criteria.

Alternatively, other countries utilize a scheme where the condenser neck is rigidly connected to the turbine, thereby eliminating the vacuum load on the deck. This scheme permits the use of much more slender columns than can be utilized for U.S. installations.

Since the conventional foundation is also low-tuned, this report does not differentiate between the low-tuned and conventional foundation concepts, except for a few of the load cases presented in Chapter 3. This is because the general proportions of the current conventional foundations have evolved from the very early massive foundations, which were essentially solid blocks of concrete with cutouts for the equipment and piping, to the lighter foundations of today, which are essentially space-frame-type structures. This evolution in the physical characteristics of the foundation has been very slow, occurring over a period of 50 years. Further, the design criteria that have been developed by the manufacturers have also evolved over this same period of time, and are largely empirical in nature. The criteria have remained empirical in the United States because there have been few attempts to perform detailed machine/foundation dynamic analyses and correlate these analyses with the measured response of operating units.

The Committee believes, then, that the low-tuned foundation concept represents an extrapolation from both past and present practice, and therefore treats the concept conservatively in some parts of the report. This is not intended to imply that the Committee questions the ability of the low-tuned foundation to perform adequately; in Europe, the low-tuned concept has been used successfully since 1936. More experience is required in the United States before the same degree of confidence will exist in the low-tuned concept as in the conventional type foundation.

Finally, this report is intended as a guide to the engineers who are responsible for the design and analysis of large turbine-generator foundations. The responsibility for satisfying the manufacturer's criteria for a particular installation is left to the responsible engineer. This report is not intended to supplement or supersede any manufacturer's criteria.

Chapter 2

Turbine-Generator Equipment

The main components of a turbine-generator are the high-, intermediate-, and low-pressure turbine sections, and the generator, exciter, and condenser. (See Figure 1 for the components of a typical turbine-generator.)

Turbines are classified as tandem-compound or cross-compound units. In tandem-compound units, one or more turbine sections are connected along a single shaft to the generator. In cross-compound units, high- and low-pressure sections are mounted on separate shafts, with an independent generator for each shaft. In most cases, each shaft rotates at a different speed.

A turbine section may be classified as single-flow, double-flow, or opposed-flow. In a single-flow section, the steam expands along a single axial direction. A double-flow section contains two symmetrical steam paths; steam enters the center of the section and expands in opposite axial directions. An opposed-flow section is similar to a double-flow in that the steam expands in opposite axial directions; however, the staging is not symmetrical.

In multiple-section design, the steam expands through a high-pressure section and may or may not return to the steam generator for reheating. The steam then expands through the remaining sections, eventually exhausting to the condenser.

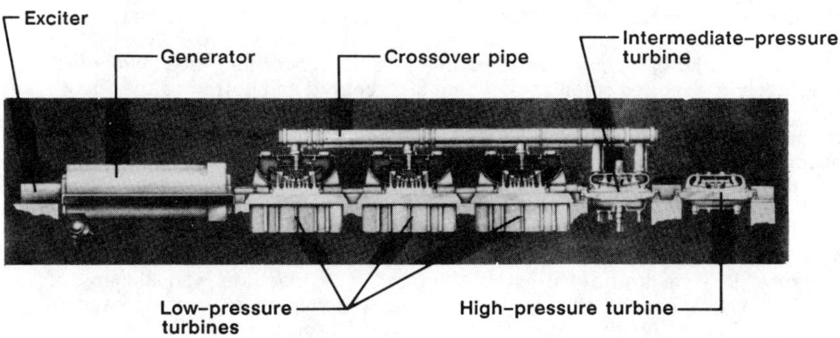

6627-1 2-86 KPS

Figure 1. Components of a Typical Turbine-Generator Set

The turbine shafts are connected to each other and to the generator rotor shaft by solid or shrunk-on couplings. Rotor field coils wound on the generator rotor are energized by an exciter, which is connected beyond the generator by a shaft that is smaller and more flexible than that of the turbine. Some generator designs use an excitation system that does not require an additional shaft. The generator rotor rotates within the stator, a circular frame that has armature coils arranged in slots in its inner circumference. An alternating electric current is induced in the armature by the rotating electromagnetic field produced by the rotor. The resistance to the flow of current in the external circuit requires the application of force to keep the electromagnetic field (and hence the rotor) moving. This force is provided by the steam expanding through the turbine sections, and is transmitted through the couplings to the generator.

Operating speeds of turbine-generator machines presently used in the United States are generally 1800 r/min for nuclear units and 3600 r/min for fossil units. A governor is provided to control machine speed and turbine output to meet system requirements. During startup, the turbine speed may be held constant at values below the running speed to prewarm the rotor and to check the clearances between the rotor and the stationary components.

Turbine-generator shafts are supported on bearings between turbine sections. Some designs use two bearings per rotor, one at each end, while other designs use a single bearing between turbine sections. On some machines, the bearings are supported within the turbine-generator casing; on other machines, the bearings are supported on the transverse beams independent of the casings. Each shaft is carefully aligned and balanced to ensure low-vibration operation. Misalignment of the shafts may result in serious vibration problems and the initiation of cracks in couplings and other rotating parts.

The low-pressure turbine exhaust nozzle is connected to the condenser. The function of the condenser is to condense the turbine exhaust steam, which is then returned to the steam supply system in the form of condensate. A vacuum is formed inside the condensers due to the volume reduction occurring as the steam changes from the vapor to the liquid state. Condensers are cooled by circulating water, which passes through the condenser shell in thin-wall tubes to enhance heat transfer without contacting or mixing with the condensate.

The turbine, generator, exciter, and governor are mounted on manufacturer-supplied soleplates, which are grouted and bolted to the foundation in accordance with the manufacturer's specifications. The turbine-generator is designed to slide on these soleplates to relieve stress due to thermal expansion and contraction. Transverse and axial guides or keys are provided to prevent misalignment of the shaft due to this thermal movement.

On most modern machines, a system that measures the magnitude of the shaft vibrations is installed. Typical information that can be obtained is shown in Figure 2. Trip limits are established for each machine. Should any of the shaft vibrations exceed this value, the unit will trip to prevent damage to the machine. Generally, the value of the limiting absolute rotor displacement is in the neighborhood of 10 mils peak-to-peak. On other machines, the relative motion between the rotor and the bearings is measured. In this case, the limiting relative rotor displacement is in the range of 3 to 5 mils peak-to-peak.

The design and configuration of foundations for turbine-generators are based primarily on the data and criteria supplied by the manufacturers of the condenser and the turbine-generator on their drawings, which should be available to the foundation designer at the time of the layout of the foundation. Manufacturers' drawings provide information concerning machine outline and arrangement, foundation loadings, soleplate and anchor bolt locations, limiting foundation dimensions, piping and penetrations, transverse and axial anchor plate locations and design forces, condenser foot loadings and movements, equipment details, and laydown requirements.

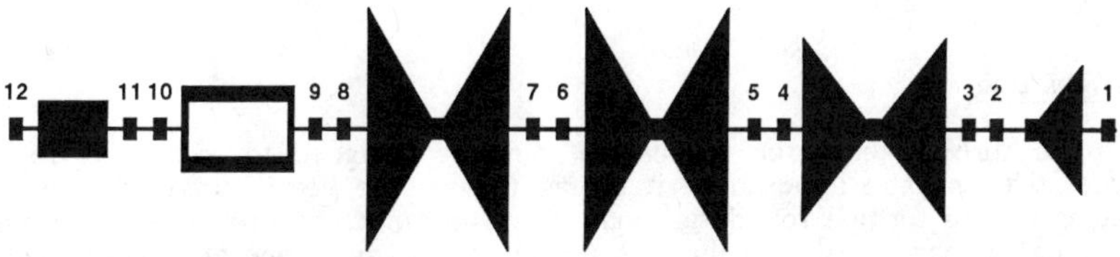

TURBINE SPEED = 3600 RPM TURBINE SPEED ROC 0.1 RPM/MIN

BEARING PARAMETERS ******** VIBRATION BAR CHARTS ********* MISCELLANEOUS **

NO.	TEMP.	VIB.	= 0 1 2 3 4 5 6 7	TRIP LIMITS (VIB.)
1	189	0.1	=	7.0
2	179	0.7	=	7.0
3	183	0.5	=	7.0
4	167	1.9	=	7.0
5	176	1.1	=	10.0
6	169	2.4	=	10.0
7	185	1.2	=	10.0
8	177	0.4	=	10.0
9	169	0.0	= *	10.0
10	147	0.1	=	10.0
11	139	2.3	=	7.0
12	147	0.9	=	7.0

6627-2 2-86 KPS

Figure 2. Example of Turbine-Generator Shaft Vibration Readout

Chapter 3

Loads

Introduction

The turbine-generator foundation must be designed to withstand all the forces that may be imposed on it during the service life of the plant. Actual loading on the turbine foundation may vary from machine to machine; however, the type of loads acting on the turbine foundation may generally be defined as follows.

Dead Load

Foundation Dead Load (DF)

The foundation dead load includes the self-weight of all members of the foundation.

Machine Dead Load (DM)

The component weight of the turbine-generator is considered the machine dead load. The turbine manufacturer provides this load as soleplate and standard-plate loads on the manufacturer's loading table.

Dead Load from Other Equipment and Piping on the Foundation (DE)

The foundation may support other assemblies, such as turbine stop and control valves, interceptor valve, boiler feed pump, main steamline hangers, etc. Additional dead loads from these assemblies, which are not included in the subsection on machine dead load, must be considered. For the dead load of the condenser, refer to the subsection on condenser dead load.

Live Load (LL)

The live load includes the loads that vary in intensity and/or occurrence. The live load from galleries and floor slabs supported by the turbine foundation should be a minimum of 200 psf (10 kPa) or as required by any local code requirements. Maintenance loads, such as laydown loads, are also considered a live load.

Condenser Load

The type of connection between the turbine and the condenser and the method of supporting the condenser at its base determine the manner in which the condenser load is transmitted to the foundation. Two common methods of supporting the condenser (see Figure 3) are as follows.

- The bottom of the condenser is mounted on rigid supports, and an expansion point is placed between the condenser and the turbine exhaust nozzle to relieve thermal forces and variations in the condenser load.
- The bottom of the condenser is mounted on springs, while the top is connected rigidly to the turbine exhaust nozzle. The springs can be adjusted to transfer specified maximum and minimum loads to the turbine exhaust nozzle. They can also be adjusted to compensate for load eccentricity, such as from the circulating water pressure load.

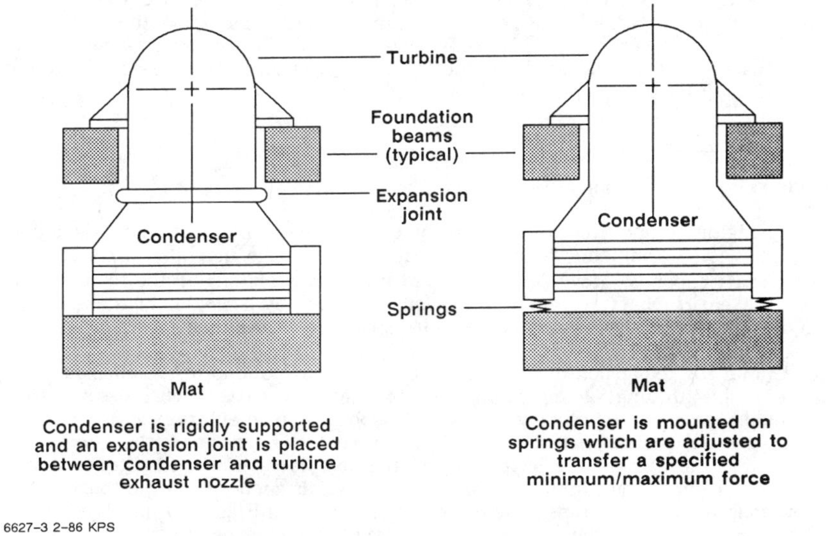

Figure 3. Major Methods of Supporting Condenser

Condenser Dead Load (CD)

When the bottom of the condenser is mounted on rigid supports, the entire dead weight of the condenser is transmitted to the mat of the turbine foundation. When the condenser is welded to the exhaust nozzle and supported on springs, the proportion of the condenser dead load distributed between the deck level and the mat level depends upon the stiffness and initial setting of the springs supporting the condenser. The loading is generally provided by the turbine and/or condenser manufacturer.

Condenser Vacuum Load (CV)

When an expansion joint is provided between the condenser and the turbine exhaust nozzle, the difference between the atmospheric pressure on the casing of the turbine and the vacuum pressure inside the condenser results in a force on the turbine. This vacuum load can be as large as several times the weight of the condenser. Typically, one condenser unit is provided for each low-pressure turbine, and its vacuum load is transmitted to the foundation through the turbine soleplates. The exact distribution of this loading is provided by the turbine manufacturer.

For the case where the condenser is rigidly attached to the turbine exhaust nozzle, no vacuum load is transmitted to the deck.

Normal Torque Load (QN)

The steam forces in each turbine section impose a torque on the stationary casings in the opposite direction from the rotation of the rotor. The normal operational torque on the generator stator is in the same direction as the rotation of the rotor. The magnitude of the torque depends on the rotational speed and the power output of the turbine section. The turbine manufacturer tabulates specific torque loads on the mechanical outline drawing as equivalent vertical loads on the soleplates.

Thermal Loading

Machine Expansion–Contraction (TM)

Changing temperatures of the turbine and generator cause expansions and contractions, forcing the various parts to slide. As the machine heats up, the entire shaft expands. However, it does not impart any loading on the foundation, since the entire shaft system is fixed longitudinally by a single thrust bearing, and the shaft slides freely across the well-lubricated journal bearings.

Unlike the expansion of the shaft, it is the heat buildup in the casings that imposes the thermal loading on the foundation. The high-pressure and intermediate-pressure turbine casings are supported on soleplates located on the top of the transverse beams. The low-pressure turbine casings, the generator casing, and the exciter are supported on the soleplates of the longitudinal and transverse beams. An example of where the casing is anchored to the foundation in the longitudinal and transverse directions is shown in Figures 4 and 5. During the thermal transient, the casing expands from the anchor points, thus producing the frictional loads. These thermal loads do not impose a net resultant force on the foundation, since the forces on any soleplate are balanced by equal and opposite forces on the anchors or on other soleplates.

The calculation of the exact magnitude of the thermal loading is very difficult, as it depends on a number of factors, such as the distance between anchor points, the condition of the sliding surface, and the magnitude of the vertical load on each soleplate. However, the following method, which utilizes simplifying assumptions, may be used in lieu of a rigorous analysis.

The total thermal loading is a vectorial sum of the longitudinal expansion loads parallel to the shaft axis and the transverse expansion loads perpendicular to the shaft axis (i.e., vectorial sum of the loads indicated in Figures 4 and 5).

The magnitude of the thermal loading under any soleplate may be calculated by:

Force = (Coefficient of friction) (Sum of machine dead load, condenser load acting through soleplate, normal torque load, and piping load) (3-1)

Thermal Loading
(Transverse loading due to machine expansion) (Startup condition)
(Transverse loading is reversed for machine contraction) (Shutdown condition)

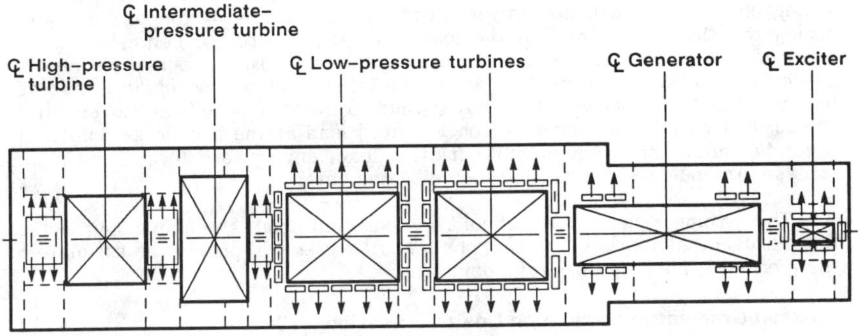

≡ Typical location at which turbine casing is anchored in the transverse direction
Transverse force = Coefficient of friction x Machine DL on soleplate

6627-4 2-86 KPS

Figure 4. Turbine-Generator Plan at Operating Floor (Transverse)

Thermal Loading
(Longitudinal loading due to machine expansion) (Startup condition)
(Longitudinal loading is reversed for machine contraction) (Shutdown condition)

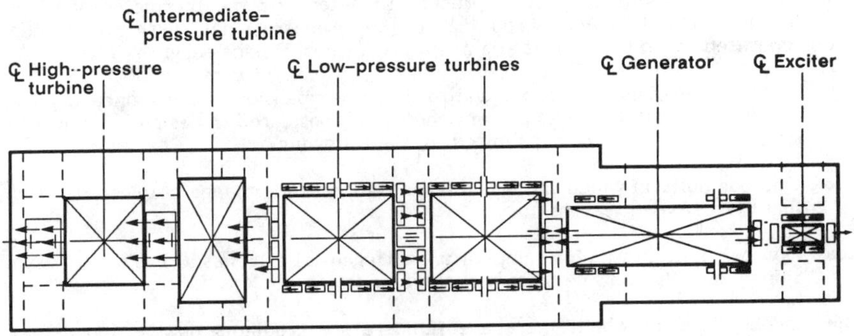

|| Typical location at which turbine casing is anchored in the longitudinal direction
Longitudinal force = Coefficient of friction x Machine DL on soleplate

6627-5 2-86 KPS

Figure 5. Turbine-Generator Plan at Operating Floor (Longitudinal)

The coefficient of friction varies from 0.2 to 0.5. The turbine manufacturer must be contacted for the actual coefficient to be used.

During the startup condition, the direction of the transverse expansion loads acting on the deck will be outward from the centerline of the turbine, while during the shutdown condition the loads will act toward the centerline of the turbine. In the case of the longitudinal expansion loads, an approximation may be made as to the direction of forces on the soleplate, and any unbalanced forces between the two adjacent anchorages which prevent longitudinal movement of the turbine may be applied as a concentrated load at the anchorage points. In order to obtain the worst possible friction load, any upward torque and piping forces that would reduce friction should be neglected.

The turbine manufacturer should be consulted about the above approach to thermal load calculations. Often, the manufacturer will provide an estimate of the thermal loading at anchorage points.

Thermal Gradient in Foundation Due to Operation (TF)

The stresses and deflections caused by the thermal gradient within the foundation due to the elevated temperatures of surfaces near the equipment are generally not considered in design. The additional deflections due to thermal expansion of the foundation may be significant; however, quantitative evaluations of these deflections are not made. The expected thermal deflection at various bearings is estimated by the manufacturer, based on past field measurements on existing units. The machine erector then compensates for the expected deflections during initial alignment. Successful past performance of turbine-generator foundations provides ample justification for such an approach. It should, however, be noted that much of the U.S. experience on thermal expansion is based on the conventional concrete foundation, and caution is necessary when low-tuned concrete or steel foundations are used. For these types of foundations it may be prudent to determine the thermal expansion; deflections should be measured during the preoperational testing and then compensated for in subsequent alignment operations, if necessary.

The stresses and deflections due to thermal expansion of the foundation due to any environmental effects present should be considered in design. Applicable situations include, but are not limited to, the following.

- Outdoor units in which there is a large temperature change between day and night or between seasons.
- Outdoor units in which a temperature differential exists as a result of the sun shining on only one side or part of the foundation.
- Indoor units in which there are temperature gradients on the foundation resulting from cold or hot air blowing over parts of the foundation.

Loading Due to Shrinkage and Creep (SC)

In reinforced concrete foundations, the time-dependent deflection at bearing locations due to creep and shrinkage can be two or more times as great as the elastic deformation upon application of the load. However, these additional deflections are generally not considered in the design for the following reasons.

- For a typical foundation, the initial machine alignment may occur 2 to 3 years after foundation construction. By this time, most of the basic creep and shrinkage have taken place. Also, the rate of creep and shrinkage deformation at the time of alignment are well below the initial rate as shown in Table 1; with a 3- to 5-year period between successive shaft alignments, the incremental values are insignificant.

- The foundation proportions are such that regardless of the absolute values of the deflections, the differential deflections between bearings are small.

Successful past performance of turbine-generator foundations provides ample validity to the above arguments. It should be noted, however, that much of the experience in the United States is based on the conventional concrete foundation, and caution is necessary when low-tuned concrete foundations are used.

Table 1

Incremental Creep and Shrinkage Deformation as a Ratio of Various Base Values

Days	Years					
28	1	2	3	4	5	20
1.00	1.75	1.96	2.08	2.16	2.23	2.64
-	1.00	1.12	1.19	1.23	1.27	1.51
-	-	1.00	1.06	1.11	1.14	1.35
-	-	-	1.00	1.04	1.07	1.27
-	-	-	-	1.00	1.03	1.22
-	-	-	-	-	1.00	1.18
-	-	-	-	-	-	1.00

Piping and Valve Reactions (Excluding DL) (TP)

Pipe forces result from cold springing, fluid dynamics, seismic forces, and thermal expansion. Pipe forces are applied at the following locations.

Turbine Casing Pipe Forces

Piping connected to the turbine casing may be cold sprung into position so that when high-temperature steam heats the system, the stresses will be relieved. Loads created by the connections should be investigated to determine if they are significant. This may be accomplished by applying the maximum permissible forces and moments that the turbine manufacturer specifies.

Maximum allowable loads are specified by the turbine-generator manufacturer to prevent distortion and/or overturning of the turbine components. The turbine casing may be assumed to be rigid and the forces then calculated at the support points on the foundation. The types of piping that generate most of the loads are:

- main steam inlet piping,
- reheat steam piping (hot and cold), and
- extraction steam piping.

Pipe Loading from Equipment Attached to the Foundation

The foundation may support such assemblies as the turbine or control valves, interceptor valve, turbine steam chest, etc. Positioning and aligning piping for this equipment creates initial erection forces. The turbine manufacturer's piping is then fitted and welded, and the assembly is anchored to the foundation. The remainder of the steam inlet pipes are then welded to the assembly inlet connections. Different forces are created due to thermal expansion during operation. Erection forces and static and dynamic operating forces should be investigated to determine if they are significant. The loading due to quick closing of the stop valve attached to the foundation is usually significant.

The turbine-generator manufacturer's piping forces are used whenever specified. These loads are the maximum allowable loads on the soleplates. The direction and point of application of these loads are given. However, if actual loads are known, they should be used. Actual forces would include the turbine-generator manufacturer's computed loads plus the purchaser's computed loads. The actual load cannot exceed the maximum load specified by the turbine-generator manufacturer.

If the turbine manufacturer does not supply piping forces at the soleplates, the designer should obtain the turbine manufacturer's maximum allowable forces at the pipe connections and then calculate forces at support points to determine if they are significant. These loads may be used during hot or cold conditions.

Normal Machine Unbalance Load (NB)

It is theoretically possible to balance the turbine-generator rotor to eliminate unbalance forces during rotation; in practice, however, some unbalance will always exist. Its magnitude depends on factors determined by design, manufacturing, installation, and maintenance procedures. These factors may include an axis of rotation which does not pass through the center of mass of the rotor, deflection of the shaft due to gravity, uneven thermal expansion, misalignment during installation, and/or corrosion or wear of moving parts. The cumulative effect of these factors is the presence of unbalance forces that are synchronous with the shaft rotational speed. These forces are transmitted to the foundation through the shaft bearings.

The effect of the normal machine unbalance loads on the turbine-generator foundation can be evaluated by a dynamic analysis or by a pseudodynamic analysis. Dynamic analysis is recommended for low-tuned foundations whereas pseudodynamic analysis has been the industry practice for conventional foundations. The normal machine unbalance load is specified as a dynamic forcing function when a dynamic analysis of the foundation is to be performed and it is specified as a pseudodynamic load when static analysis is performed.

Dynamic Forcing Function

For the purpose of defining the normal machine unbalance dynamic load, the turbine-generator rotor can be approximated by a multiply supported shaft with a rotating disc corresponding to each of the turbine stages and the generator as shown in Figure 6. The forcing function F(t) corresponding to each of the turbine stages or the generator is:

$$F(t) = M_i \frac{G\Omega^2}{\omega} \sin(\Omega t + \alpha_i) \tag{3-2}$$

where:

M_i = the mass of the rotating mass i;

G = $e\omega$, a measure of the balance quality grade of the rotor;

e = the rotating mass eccentricity, which equals the distance between the axis of rotation and the mass center of the rotating mass (see Figure 7);

ω = the machine design operating speed in radians/second;

Ω = the rotational speed in radians/second for which the unbalance force is being computed; and

α_i = the relative phase angle of the unbalance for rotating mass i.

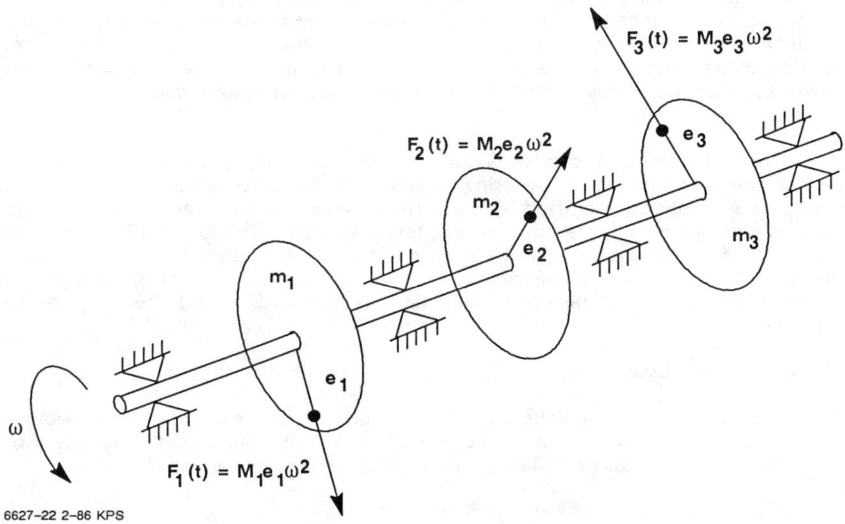

Figure 6. Unbalance Forces Along Shaft with Multiple Supports

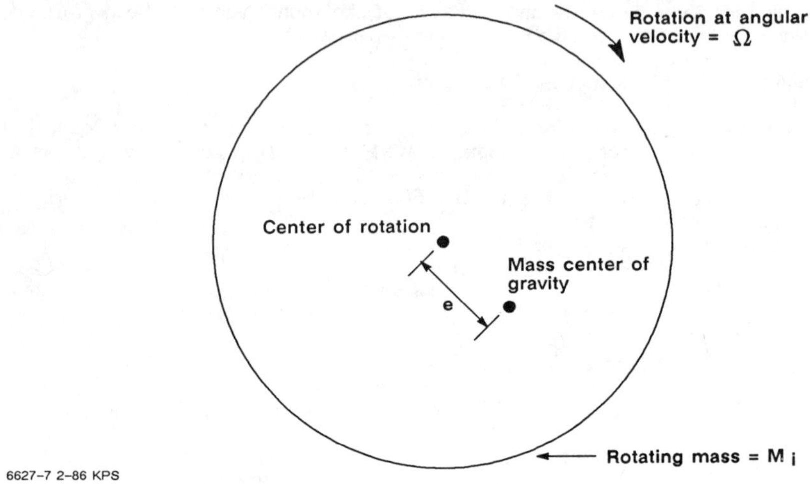

Figure 7. Rotating Eccentric Mass

For turbine-generator sets, the balance quality grade per ISO 1940 (18) is G 2,5. This corresponds to a G value of 0.1 in/s (2.5 mm/s). The relative phase angle α_i of the unbalance forcing function corresponding to each stage of the turbine and generator is random and therefore unknown at the design stage; however, its random nature should be considered in the determination of the probable maximum response of the turbine-foundation system due to the normal machine unbalance.

The 0.1 in/s (2.5 mm/s) G value corresponds to a minimum operating unbalance, and the actual operating unbalance is likely to be larger. The design unbalance should be specified by the turbine-generator manufacturer. Based on discussions presented later in this chapter, a G value of 0.20 in/s (5 mm/s) may be used when the manufacturer's information is not available. For the design unbalance, the maximum bearing cap deflection should be limited to the trip limit (~10 mils peak-to-peak) for machine speeds from 20% to 120% of operating speed.

Pseudodynamic Load

The normal machine unbalance forcing function is a dynamic load; however, for conventional foundations, the effects on the foundation response are generally evaluated using a static analysis and pseudodynamic loads.

The magnitudes of these pseudodynamic loads based on past practice can be summarized as follows.

- Vertical pseudodynamic downward loads on the foundation soleplates with the following magnitude:

  25% of DM for 1800 r/min machines, and

  50% of DM for 3600 r/min machines,

  where DM is the turbine-generator machine dead load.

- Transverse horizontal pseudodynamic loads equal in magnitude to 50% of the turbine-generator machine dead load supported by each pier, applied at the shaft centerline elevation in the direction perpendicular to the shaft axis.

- Longitudinal horizontal pseudodynamic loads equal in magnitude to 50% of the turbine-generator machine dead load supported by each pier, applied at the shaft centerline elevation in the direction parallel to the shaft axis.

Correlation of Dynamic and Pseudodynamic Loads

In the past, the effect of normal machine unbalance has been considered using the pseudodynamic loads described in the previous subsection. These pseudodynamic loads can be compared to those obtained using the dynamic loads defined earlier to provide a benchmark for the design G value.

The magnitudes of the pseudodynamic loads can be computed based on the dynamic load magnitude and the dynamic load factor (DLF). The pseudodynamic load for the dynamic load defined by Equation 3-2 is

$$F_{static} = (DLF)\, M \,\frac{G\Omega^2}{\omega} \qquad (3\text{-}3)$$

DLF is defined as the ratio of the dynamic response to the static response and is given by:

$$DLF = \frac{1}{\left[\left(1 - \left\{\frac{\Omega}{f_r}\right\}^2\right)^2 + \left(2\xi\,\frac{\Omega}{f_r}\right)^2\right]^{1/2}} \qquad (3\text{-}4)$$

where:

Ω = angular velocity (rad/s) of the rotor for which the DLF is being computed,

f_r = critical natural frequency (or frequencies) of the rotor foundation system, and

ξ = critical damping ratio (usually 0.02 for concrete foundations).

Table 2 presents the maximum pseudodynamic load for a G value of 0.1 in/s (2.5 mm/s) for 3600-r/min and 1800-r/min machines at machine operating speed and at rotor critical and foundation critical speeds. For a 3600-r/min machine, the computed maximum pseudodynamic load is 0.244 DM, compared to 0.5 DM used in conventional design. For 1800-r/min machines, the computed maximum

Table 2

Machine Unbalance Pseudodynamic Load for G = 0.1 in/s (2.5 mm/s)

Machine Speed	Pseudodynamic Load: Horizontal	Pseudodynamic Load: Vertical
3600-r/min Machines		
@ Operating speed 3600 r/min	0.012 DM	0.012 DM
@ Shaft critical speed 2400 r/min	0.244 DM	0.244 DM
@ Foundation critical speed H = 1200 r/min V = 2400 r/min	0.122 DM	0.244 DM
1800-r/min Machines		
@ Operating speed 1800 r/min	0.006 DM	0.183 DM
@ Shaft critical speed 1200 r/min	0.122 DM	0.122 DM
@ Foundation critical speed H = 1200 r/min V = 4800 r/min	0.122 DM	0.183 DM

Note: DM = Dead weight of the turbine-generator components.

pseudodynamic load in the horizontal direction is 0.122 DM compared to 0.5 DM used in conventional design. In the vertical direction, the computed maximum pseudodynamic force is 0.183 DM compared to 0.25 DM used in conventional design. For these calculations, it was assumed that the rotor weight is 15% of the total machine weight (DM).

Based on the above, a design G value of 0.2 in/s (5 mm/s) for dynamic analysis would lead to results consistent with past practice.

Seismic Load (EE)

The forces calculated in accordance with the Uniform Building Code (UBC) (17) provide minimum standards for design criteria to make structures earthquake resistive. The lateral forces are applied separately to the foundation in the longitudinal and transverse direction. The UBC gives the total lateral seismic force as:

$$V = ZICSKW \tag{3-5}$$

where:

Z = a numerical coefficient dependent upon the seismic zone and determined from appropriate zone maps in the UBC.

I = the importance factor, which is assigned a value which varies from 1.0 to 1.5.

C = a numerical factor related to the amplification factor of the structure. The value of C need not exceed 0.12.

S = a numerical factor to account for soil-structure interaction effects. It need not exceed 1.5. Moreover, the product CS need not exceed 0.14. To minimize detailed analyses use CS = 0.14.

K = the building coefficient, which is assigned a value of 1.0.

The seismic excitation can also be specified in terms of ground response spectra if a dynamic analysis is performed to determine the seismic response.

Generator Emergency Torque (QE)

Of all the short-circuit faults that can occur, a line-to-line short circuit at the generator terminals causes the most severe loading of the turbine-generator foundation. Such a fault occurs when any two of the three generator phase terminals are shorted. The calculation of the maximum generator air-gap torque during symmetrical (three-phase) and unsymmetrical (line-to-line and line-to-ground) terminal short circuits is normally performed assuming no electrical damping in order to obtain the greatest possible forces that can be transmitted to the foundation under different fault conditions. The results show that the maximum torque resulting from a line-to-line short circuit is about 25% greater than that caused by a single terminal-to-ground fault and roughly 30% higher than that associated with a symmetrical three-phase fault at the terminals of the same generator.

The loading due to generator short circuits is generally provided as either a forcing function or as equivalent static forces. The use of equivalent static loads for the maximum short-circuit torque assumes that the foundation is infinitely rigid and consequently must directly absorb the full impact of the severe shock forces. Since this assumption may result in overdesigning the foundation, the more realistic approach of a dynamic analysis on the basis of the short-circuit torque time function is generally preferred.

Forcing Function

The synchronous-machine equation, which describes the air-gap torque from the instant a line-to-line short circuit begins, is shown at the top of Figure 8 as a graphical representation of the three constituent elements: the 60- and 120-Hz oscillating torque components and the unipolar torque component. Each generator design is characterized by different coefficient values (A, B, and C). The generator supplier will provide data about these coefficients. The resultant alternating torque, depicted as a trace in the lower half of Figure 8, is a summation of the three components. It attains its peak value ($T_{s.c.max}$) when the angular rotation of the rotor equals approximately 2.1 radians, which occurs about 5.5 ms after the incidence of the fault.

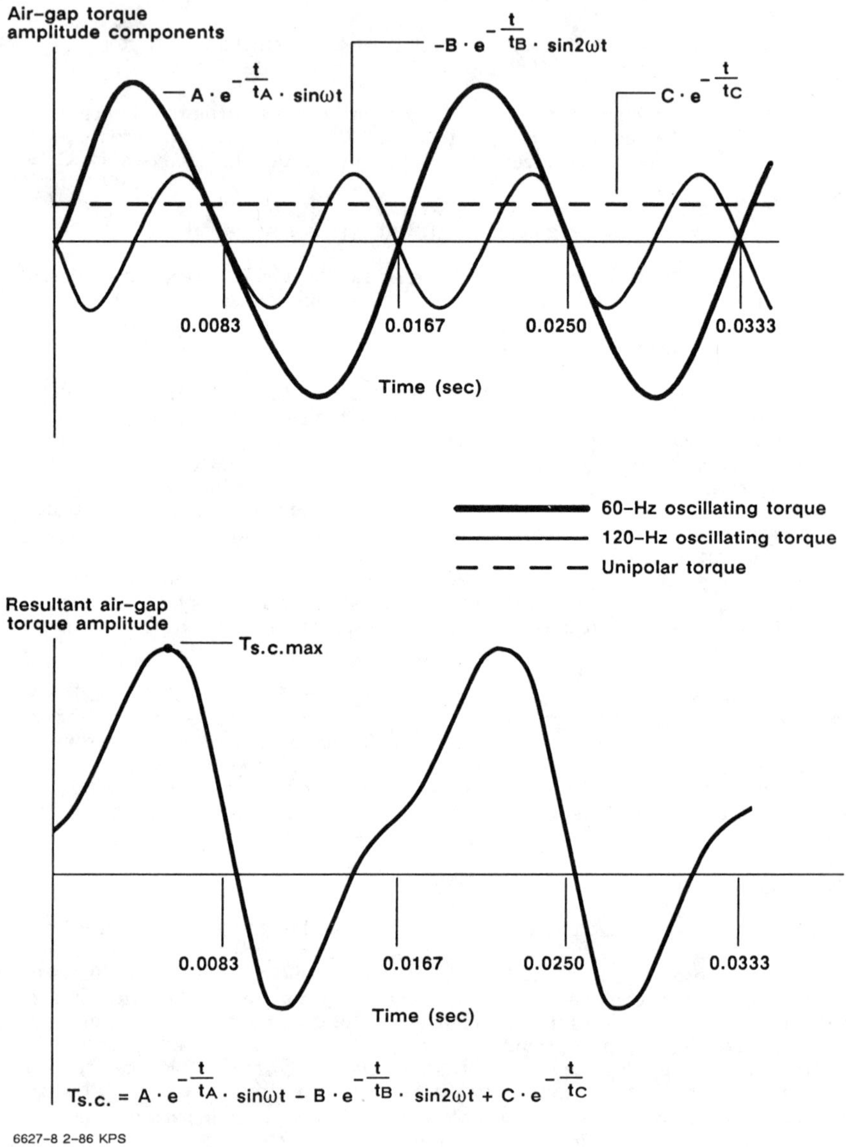

Figure 8. Air-Gap Torque Due To Generator Short Circuit

In view of the very severe transient nature of the maximum short-circuit loading of the foundation in the vicinity of the generator, the foundation designer should perform an appropriate dynamic analysis of this abnormal load case.

Equivalent Static Forces

Turbine-generator suppliers can also express the short-circuit loading as equivalent static forces that act vertically at the soleplate locations where the generator is mounted on the foundation. Such foundation loads are derived from the same terminal-to-terminal short-circuit torque equation by correctly allocating the peak value ($T_{s.c.max}$) to the individual soleplate pairs and dividing these parts of the total torque by the corresponding operative moment arms.

Load Due To Out-of-Phase Synchronization (MS)

Out-of-phase synchronization means that the wave form of the voltage being produced by the generator does not have the same time relationship with the system voltage phaser at the instant the generator is connected to the system. Faulty synchronization can impose severe alternating forces on the foundation in the vicinity of the generator. The maximum air-gap torque resulting from worst-case malsynchronization (120 electrical degrees out of phase) is a function not only of the generator design parameters, but also of the characteristics of the electrical transmission system to which the generator is connected.

If the sum of the main stepup transformer reactance and the system reactance equals the subtransient reactance of the generator, the 120° out-of-phase synchronization causes a shock force on the foundation roughly equivalent to the maximum line-to-line short-circuit torque ($T_{s.c.max}$). In the case of a relatively small transformer reactance and a strong system, the worst-case malsynchronization air-gap torque can lie in the range of 10% to 35% greater than the $T_{s.c.max}$ value. Even if the system and transformer reactances are both considered to be zero (i.e., the generator is connected directly to an infinitely strong system), the theoretical maximum malsynchronization torque is only about double the $T_{s.c.max}$ amplitude.

The turbine manufacturer will specify the magnitude of this loading in the form of a forcing function or equivalent static forces should he desire to have any part of the foundation designed for this loading. However, generators are not designed for any electrical disturbances that are more severe than terminal short circuits. Due to the extremely low probability of faulty synchronization occurring with a sufficiently large angular discrepancy between the generator and system voltage phasers to result in an air-gap torque peak which exceeds the $T_{s.c.max}$ value, it is an acceptable practice not to consider the exceptional case of gross malsynchronization as a criterion for foundation design. This assumes that the effects of worst-case out-of-phase synchronization can be covered by the overall foundation design margins.

Load Due To Bowed Rotor (AB)

A bowed rotor can impose large dynamic forces on the turbine-generator foundation. The bowed condition of the rotor will create unbalance forces which are transmitted through the machine bearings to the soleplates. The magnitude of the forces will vary with the square of the speed, the weight of the rotor, and the amount of eccentricity in the rotor.

A bowed rotor can occur in any turbine section and can be the result of the following conditions:

- unusually severe packing rub,
- water induction, and
- failure to put the rotor on a turning gear when the machine is shut down.

The first condition will cause a differential temperature in the rotor, which will cause the rotor to bend, resulting in the unbalance.

The second condition, water induction, can occur when a slug of water enters the turbine, causing a differential temperature and bending of the rotor shaft.

The third condition can be caused by improper operation or system failure and not placing the rotor on turning gear operation while cooling down.

The largest bowed rotor response occurs at the first critical speed for the rotor. The length of time that it takes for the turbine rotor to pass through the critical speed is a relatively short period while going on line. However, the time is much longer when the machine is being taken off line and the rotor coasts through the resonant speed.

The probability of a bowed rotor is difficult to estimate; however, it is likely that some degree of bowing will occur during the life of the unit. Since this is a condition that usually requires turbine-generator shutdown, it will exist only for the time required for the rotor to coast down to rest. Therefore, it is sufficient to assure that stresses in the foundation are low enough to preclude any permanent damage to the structure during the coastdown period.

The magnitude and location of the forces due to a bowed rotor are supplied by the manufacturer of the turbine in question and will of necessity be based on the particular assumptions made by that manufacturer.

The forces due to a bowed rotor can be calculated with a relationship similar to that for an unbalanced rotor as follows:

$$F = Me\omega^2 \sin \omega t \tag{3-6}$$

where:

F = force due to bowed rotor,

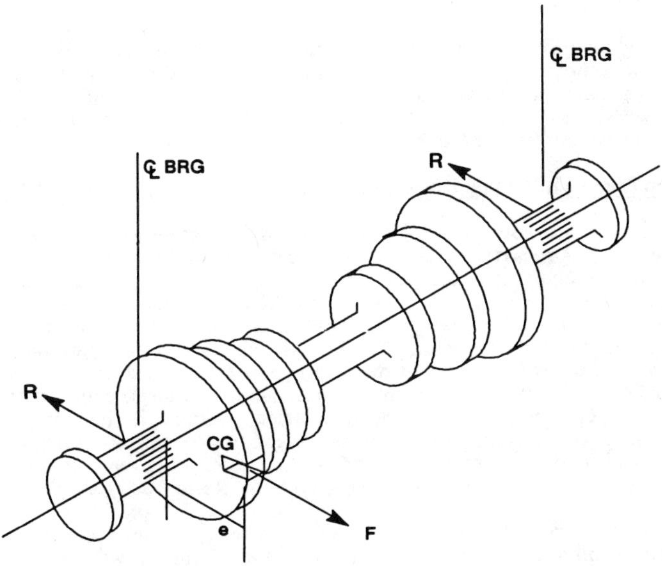

Figure 9. Load Due To a Missing Rotor Blade

M = mass of rotor,

e = assumed rotor eccentricity, and

ω = critical circular frequency of the shaft or foundation.

The loading will be provided in the form of a sinusoidal forcing function for the dynamic analysis or equivalent static loads for a simplified analysis of the foundation.

Some turbine manufacturers assume that a bowed rotor is the worst case of turbine emergency for the high- and intermediate-pressure sections. Other turbine manufacturers assume that the loss of turbine blade (see next section) is the worst turbine emergency and do not consider a bowed rotor emergency load in their summation of loads for the turbine.

Load Due To a Missing Rotor Blade (AM)

A turbine rotor must be balanced dynamically within practical limits in order to ensure satisfactory turbine operation and cause no adverse effects on the turbine equipment and/or foundation. However, a more severe unbalance can occur while the unit is in operation. This emergency unbalance condition is predicated on the unlikely event that a last-row blade in a low-pressure rotor breaks loose from the rotor. The loss of this blade, which can range in size from about 20 inches (500 mm) to over 40 inches (1000 mm) in length, could cause a significant unbalance in the rotor; consequently, a large dynamic force would be imposed on the rotor/bearing/foundation system. The magnitude of this unbalance is a function of the blade weight, its center of gravity with respect to the rotor, and the rotational speed of the rotor. This force is transferred to the foundation through the rotor bearing system. Refer to Figure 9.

As this condition can be postulated to occur in any of the several rows of last-row blades, a separate analysis should be made with a single unbalance equivalent to the loss of one last-row blade applied to the mass point corresponding to each of the last-blade rows in each low-pressure turbine. Since this is an emergency condition that will require turbine-generator shutdown, it will exist only for the time required for the rotor to coast down to rest. Therefore, it is sufficient to assure that the stresses in the foundation are low enough to preclude any permanent damage to the structure during the coastdown period.

The magnitude and location of the forces due to the loss or breakage of a last-row blade are supplied by the turbine manufacturer. The loading is provided in the form of a sinusoidal forcing function for dynamic analysis or equivalent static loads for simplified analysis of the foundation.

Chapter 4

Foundation Geometry

This chapter presents a description of and criteria for the layout and initial sizing of a conventional concrete turbine-generator foundation, including the overall foundation, operating deck, columns, mat, and intermediate platform. Also presented are the results of a nationwide survey of turbine foundation designs.

Overall Foundation Proportions

The turbine foundation is located within the main turbine building, which houses other equipment and piping necessary for power generation. The deck is kept structurally isolated from the surrounding building by a gap of 1 inch (25 mm) or more at all levels; the gap may be filled with a compressible material at the floor levels. This isolation is intended to prevent the transfer of machine and deck vibrations to the adjoining structure.

The turbine-generator foundation is constructed of monolithic reinforced concrete consisting of a foundation mat, columns, and the deck. Partial mezzanine floors and cross-walls adjacent to the condenser shells may also be provided, if required. Except for cross-compound units, the foundation is normally symmetrical with respect to the vertical plane passing through the longitudinal axis of the machine. The conventional foundation (excluding the mat) normally weighs three to five times the weight of the machine it supports. Figures 10 through 12 show a general layout of a typical large turbine-generator foundation.

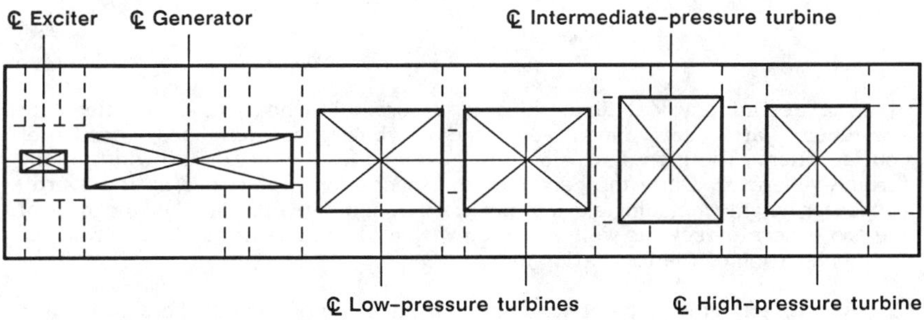

Figure 10. Operating Floor Plan - Turbine-Generator Foundation

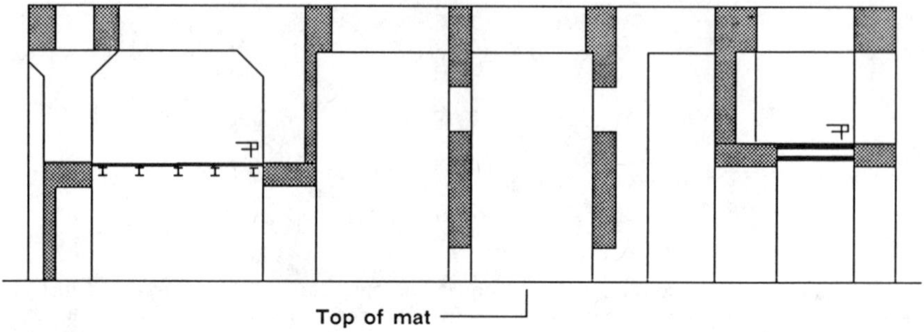

6627-12 2-86 KPS

Figure 11. Longitudinal Section - Turbine-Generator Foundation

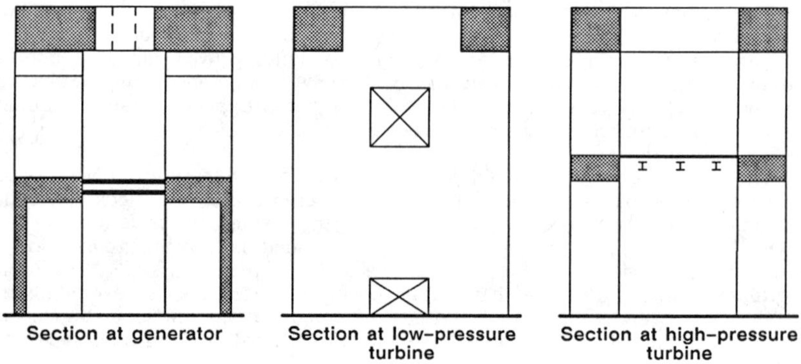

6627-13 2-86 KPS

Figure 12. Transverse Sections - Turbine-Generator Foundation

The overall physical dimensions of a turbine foundation, conventional or low-tuned, are determined by the turbine-generator and condenser configuration. The plan dimensions are governed by the length and width of the turbine-generator, while the height of the foundation is governed by the overall condenser height (including exhaust neck, shell, and hotwell) plus the thickness of the top concrete deck, as well as the elevation of the operating floor in relation to the elevation of the foundation mat.

As with any other type of structure, initial sizing of the members is required before any static or dynamic analysis is performed to satisfy the strength, rigidity, and frequency criteria. Certain empirical rules that have been established based on past experience play an important role in initial layout and preliminary design. The various member stiffness ratios specified in the succeeding discussion are based on such empirical rules.

Operating Deck Proportions

The top horizontal frame of the foundation, called the deck, supports the turbine-generator unit. This deck is composed of reinforced concrete beams which are both parallel and perpendicular to the shaft of the machine (longitudinal and transverse beams).

The layout of the deck is the first step in the foundation design. The plan shows the width and total length of the deck. The dimensions of the clear openings shown on this layout for the turbine sections and the generator correspond exactly to the opening dimensions given on the manufacturer's machine outline drawings. The soleplates, as shown on the manufacturer's drawings, are then located on the top deck plan. The overall length of the foundation is determined by providing an appropriate edge distance between the end of the soleplate and the edge of the concrete at the ends of the deck.

The width of the deck is determined by the width of the openings and the width of the longitudinal beams. The opening for the turbine exhaust is wider than the opening for the generator at the other end of the foundation. The width of the foundation at the generator end may therefore be proportionately reduced for economy and to achieve a reasonable width-to-depth ratio for the longitudinal generator beams. The transition for width reduction can either be abrupt or tapered at 45°.

The clear span of the longitudinal beams supporting the turbine is determined by the size of the condensers. The ratio of clear span to the depth of the turbine girders normally ranges from 1.5 to 3.0. The depth-to-width ratio of the girders generally ranges from 1.0 to 3.0. Alternatively, a uniform thickness for the entire operating deck may be provided. The beam width, in this case, is established such that the beam cross section has sufficient rigidity and strength and provides the required clearances and openings.

The generator is supported by soleplates located on longitudinal beams. The clear span-to-depth ratio of these beams is approximately 3 to 4. The top slab required at the generator end to support the exciter is from 4 feet to 5 feet (1.2 to 1.5 meters) thick.

In accordance with the manufacturer's drawings, embedded pipe sleeves for the oil pipes to pass through and embedded plates to resist frictional forces are located in the operating deck.

Vertical dimensions from the centerline of the shaft to the machine baseline and the turbine exhaust-condenser connection are given on the machine outline drawings supplied by the manufacturer. This information, along with the condenser tube bundle dimensions and, if required, the arrangement of feedwater heaters in the condenser neck, allow the foundation height and turbine building floor elevations to be determined. The baseline of the machine is kept either at the operating deck elevation or slightly higher to provide a drainage slope in the top deck without interfering with the lagging.

The concrete for the foundation deck is placed 2 to 4 inches (50 to 100 mm) below the final elevation in order to provide a finish floor of concrete to obtain a better appearance for the deck. The thickness of the machine soleplates and their relative elevation with respect to the machine baseline are given on the machine outline drawings. The elevation of the foundation deck concrete in the

Table 3

Operating Deck Proportions

Item No.	Beam Location	Ratio of Clear Span to Depth		Ratio of Depth to Width	
		Range	Average	Range	Average
1	Longitudinal beam supporting generator	1.44 to 4.33	2.73	0.49 to 1.75	0.88
2	Longitudinal beam supporting low-pressure turbines	1.50 to 4.20	2.67	0.56 to 1.75	1.07
3	Longitudinal beam supporting intermediate-pressure turbines	0.91 to 3.78	1.67	0.77 to 2.50	1.17
4	Longitudinal beam supporting high-pressure turbines	0.50 to 2.92	1.76	0.44 to 2.50	1.14
5	Transverse beam supporting generator/exciter (end frame)	0.87 to 4.61	2.15	0.25 to 2.30	0.82
6	Transverse beam between generator and low-pressure turbine	0.63 to 2.88	1.44	0.80 to 6.56	3.03
7	Transverse beam between two low-pressure turbines	0.83 to 3.53	1.83	0.99 to 5.36	2.82
8	Transverse beam between low-pressure and intermediate-pressure turbines	1.17 to 4.04	2.11	1.14 to 4.48	2.36
9	Transverse beam between intermediate-pressure and high-pressure or low-pressure and high-pressure turbines	0.78 to 4.74	2.06	0.54 to 5.67	2.41
10	Transverse beam supporting high-pressure turbines	1.30 to 5.67	2.69	0.63 to 4.17	1.39

Table 4

Column Proportions

Item No.	Column Location	Ratio of Unsupported Length to Area (ft^{-1})		Ratio of Total Length (top of mat to bottom of operating deck) to Area (ft^{-1})	
		Range	Average	Range	Average
1	Column supporting generator/exciter (end frame)	0.08 to 1.59	0.33	0.14 to 1.71	0.52
2	Column between generator and low-pressure turbine	0.09 to 0.92	0.29	0.20 to 1.19	0.44
3	Column between two low-pressure turbines	0.21 to 1.19	0.54	0.26 to 1.20	0.60
4	Column between low-pressure and intermediate-pressure turbines	0.15 to 1.59	0.45	0.33 to 1.59	0.64
5	Column between intermediate-pressure and high-pressure or low-pressure and high-pressure turbines	0.06 to 1.36	0.38	0.17 to 1.36	0.59
6	Column supporting high-pressure turbines only (end frame)	0.13 to 1.59	0.44	0.31 to 1.90	0.71
	All columns	0.06 to 1.59	0.38	0.14 to 1.90	0.57

Note: For SI units (m^{-1}), multiply above ratios by 3.28.

area where the soleplates are located can therefore be determined with appropriate provisions for grout under the plates (see Chapter 7).

Column Proportions

Columns can be located from the top deck plan by following the simple rules listed here:

- Column sizes are kept within the width of the beams to avoid protruding into the opening areas.
- As far as practical, compressive stresses and elastic shortening are kept uniform in all columns to help maintain the shaft alignment.
- Columns are normally placed between each section of the turbine.
- At the generator end, the columns are located as close to each other as possible under the generator in order to minimize the girder span length. To keep the columns a reasonable size, a deep cantilever off the top deck from the last column set may be provided to support the exciter (see Figure 11).

Alternatively, columns may be placed at the extreme exciter end, with an additional column placed between the exciter column and the column between the generator and the low-pressure turbine section.

Foundation Mat Proportions

Standard practice and proven performance dictate that the columns be supported on a mat. The foundation must be sufficiently rigid to preserve the shaft alignment. It should also be able to transmit accident loads to the subgrade without exceeding the allowable bearing capacity of the soil. Dowels for the walls and columns should be accommodated in the mat thickness. Normally, 5- to 6-foot-thick (1.5- to 1.8-meter-thick) mats are provided for smaller units; for larger units, thicknesses up to 8 or 9 feet (2.4 to 2.7 meters) may be provided.

Several proven rules of thumb can be used to estimate a foundation mat thickness that will provide the desired results. For mats on soil, piles, or drilled piers, the following rules apply.

- The ratio of the mat bending stiffness to the column bending stiffness in the transverse direction should be at least two. If the ratio is less than two, it cannot be assumed that the column ends are fixed.
- The mat thickness should not be less than $0.07L^{4/3}$, in which L equals the average of two adjacent clear spans (14). For mats with piles or drilled piers clustered around columns, this rule is not applicable. This thickness criterion addresses the problem of differential settlement in the mat due to nonuniform soil pressure. The above equation will result in a mat with a rigidity that justifies a uniform soil loading distribution.
- The mass of the mat should be at least equal to the mass of the operating deck plus the turbine-generator. In the past, this criterion has resulted in foundations which are stable, and it indirectly gives the mat sufficient rigidity to cause uniform soil pressure.

Although the above rules of thumb are empirical, they nonetheless offer the designer an initial value for the mat thickness. It is recommended that at least one of the criteria listed above be satisfied unless a more detailed analysis or design is undertaken. Determining which criteria, or how many, are to be satisfied is left to the judgment of the designer. The final determination of mat thickness will depend on the design moments and forces.

Intermediate Platform Proportions

Intermediate platforms may be provided, preferably at the same level as the turbine building mezzanine floor, beneath the high-pressure turbine and the generator. These platforms are preferably of concrete framing. In this case, the concrete beams are normally 3 to 4 feet (0.9 to 1.2 meters) deep, with a slab thickness approximately one-half the beam depth. Whether framed with steel or concrete, the members should be checked to ensure that no resonance occurs at operating speed.

Foundation Proportions Survey

The data relating to turbine-generator foundation member sizes for 63 existing plants designed by 11 U.S. architect-engineering firms were collected and analyzed to study the state of the art of foundation proportions. These plants, located primarily in the United States, ranged in capacity from 135 MW to 1375 MW and included both nuclear and fossil plants.

The analysis of the data revealed a wide scatter in member sizes and proportions. The results of the analysis are summarized here.

Mat Thickness

There was no relation between the mat thickness and the capacity of the plant. The mat thickness varied from 5 to 12 feet (1.5 to 3.7 meters), irrespective of the plant generating capacity.

Ratio of Foundation Weight to Machine Weight

The ratio of the weight of the turbine-generator foundation (excluding the mat) to the weight of the machine ranged from a low of 2.44 to a high of 6.31 with an average ratio of 4.03.

Operating Deck Proportions

There was a wide scatter of data for the operating deck proportions. The range and average of the ratios of clear span-to-depth and depth-to-width for the longitudinal and transverse beams of the operating deck are listed in Table 3.

Column Proportions

The range and average of the ratio of the unsupported column length to the column cross-sectional area and the ratio of the total length (top of the mat to bottom of the operating deck) to the column area are listed in Table 4.

Chapter 5

Analysis Techniques

Turbine-generators are subjected to various loads, some static, some resulting from dynamic conditions, and others resulting from shrinkage and thermal expansion and contraction. The purpose of analysis is to determine whether a proposed foundation will perform satisfactorily under the application of these loads. Two criteria must be satisfied: (1) deflection criteria, and (2) strength criteria. Deflection criteria must be satisfied to ensure that the turbine-generator alignment will be maintained. Strength criteria must be met to ensure that the foundation will not experience a structural failure.

Recommended procedures for the static and dynamic analyses, including the soil-structure interaction effects for the turbine foundation superstructure and base mat, are presented in the following sections.

Recommended Analysis for Various Loadings

The type of analysis that should be performed depends on whetner the loading is static or dynamic. However, some dynamic loads can be adequately analyzed by means of a pseudodynamic analysis, in which an equivalent static load is applied in place of the actual dynamic load. Some considerations that dictate whether a pseudodynamic analysis or an actual dynamic analysis is performed are the type of foundation used and the requirements of the turbine-generator manufacturer. A pseudodynamic analysis is generally sufficient for a conventional foundation. A low-tuned foundation usually requires that dynamic analyses be performed.

The following classification of analysis type by load (see Chapter 3 for definitions of these loads) is intended to serve only as a guide. These are general recommendations, and analyses other than those recommended should be performed if special considerations warrant.

Loads for Which a Static Analysis Is Recommended

- Dead load
- Live load
- Condenser load
- Normal torque load
- Thermal loading due to machine expansion/contraction
- Piping loading

Loads for Which a Pseudodynamic Analysis Is Recommended

- Normal machine unbalance load for machines mounted on conventional foundations
- Seismic loading based on Uniform Building Code or ANSI requirements

Loads for Which a Dynamic or Pseudodynamic Analysis Is Acceptable

- Loading due to generator short circuits
- Loading due to out-of-phase synchronization
- Loading due to bowed rotor
- Loading due to missing rotor blade

Loads for Which a Dynamic Analysis Is Recommended

- Normal machine unbalance load for machines mounted on low-tuned foundations
- Seismic loading based on the response spectrum method

Simplified Analysis

The primary objective of the simplified analysis is to determine the approximate stresses and deflections that are created in the foundation superstructure as a result of the applied loads. These models are recommended only to determine the preliminary member sizes of the foundation superstructure. Once these preliminary sizes are determined, a detailed static and dynamic analysis, as outlined in the next two sections of this chapter, should be performed.

The simplified models of the superstructure are shown in Figures 13 and 14. The superstructure is reduced to two different sets of plane frames. The first plane frame is shown in Figure 13 and represents the side view of the foundation, in which all vertical loads and all horizontal loads parallel to the axis of the turbine are applied. The columns are restrained from translation and rotation at their base. The second set of plane frames is shown in Figure 14 and represents the transverse bents. Each frame is analyzed as though it were acting independently. The loads applied to each frame reflect the actual loads acting on the transverse bent.

The stresses and deflections resulting from the loading imposed on these simplified models can be determined from a computer solution of the frame, or the solution can be determined using published solutions to frame structures as given in References 23 and 24.

The simplified models yield results which are not exact and not necessarily conservative because they do not account for the effects of mat flexibility and the interaction between adjacent bents and between the mat and soil. In addition, the simplified models should only be used to compute the response due to static or pseudodynamic loads. Dynamic analysis based on these models is not recommended because the models can only be used to estimate the fundamental natural frequencies of the foundation. All turbine foundations have participating

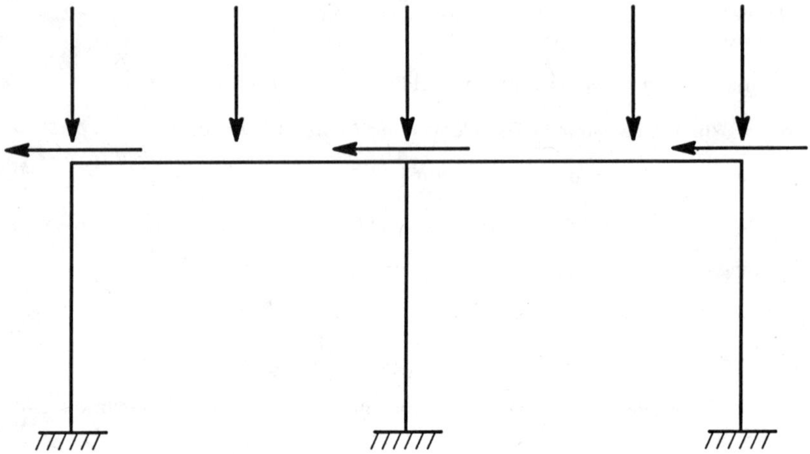

6627-14 2-86 KPS

Figure 13. Simplified Model - Longitudinal Direction

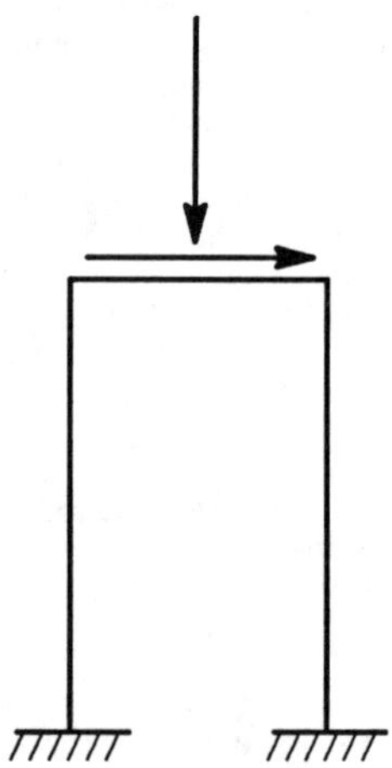

6627-15 2-86 KPS

Figure 14. Simplified Model - Transverse Direction

higher mode frequencies at or near the operating speed; therefore, a frequency check, in itself, does not ensure the adequacy of the foundation.

Detailed Static Analysis

The detailed static analysis should be performed after a preliminary analysis has been completed. This subsection describes the modeling methods, special considerations for modeling, and the computer programs used to perform the detailed static analysis.

Modeling Methods

There are two acceptable methods to model the foundation superstructure and the mat.

In the first method, the deck and columns are modeled so that the columns are fixed at their base (as shown in Figure 15). The loads are applied to the model of the superstructure to determine the foundation response. The computed reactions at the base of the columns are then used as input forces on the model of the mat.

In the second method, the deck, columns, and mat are incorporated into a single model as shown in Figure 16. This procedure yields results for the entire foundation in one analysis.

The first method is most commonly used and is recommended for the detailed static analysis. However, the second method will generally lead to smaller member forces but larger deflections. It is also more accurate.

For either of the two methods described above, the turbine foundation is modeled as a three-dimensional space frame in which the columns and beams are modeled using three-dimensional beam elements. Such a model is shown in Figures 15 and 16. Shear walls within the superstructure are modeled using plate-bending and membrane elements.

The base mat, if modeled, is modeled with plate-bending elements. The nodes of the mat lie in a horizontal plane at the middle surface of the mat. The interaction between the mat and soil is modeled as a series of horizontal and vertical springs which connect the nodes to fixed supports. References 10, 11, and 31 illustrate the formulation of these models. The soil-structure interaction effects are discussed later in this chapter.

The turbine and generator are not included as part of the static analysis model. The weight of the machine is an external load acting on the soleplates at the point where the machine interfaces with the foundation.

Special Modeling Considerations

Because of the small length-to-depth ratios of the members of a turbine foundation, a number of factors must be considered when creating the computer model of the foundation.

1. The shear deflections of the girder elements should be included in the analysis.

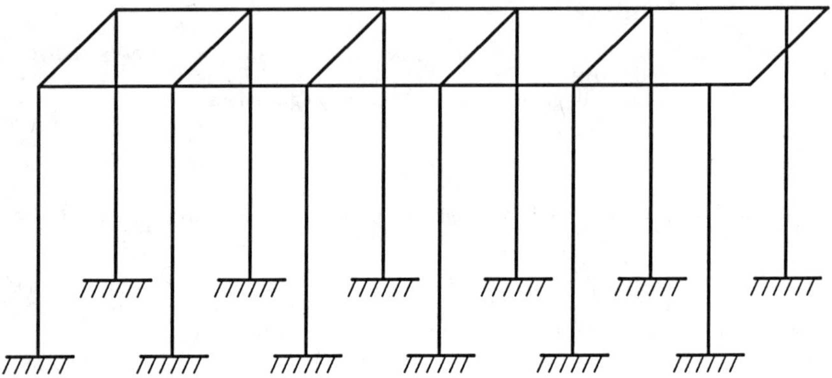

6627-16 2-86 KPS

Figure 15. Three-Dimensional Computer Model of Foundation Superstructure

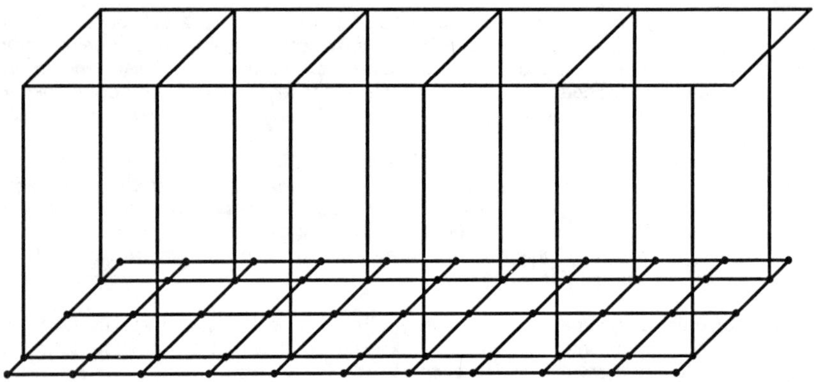

6627-17 2-86 KPS

Figure 16. Three-Dimensional Combined Model of Foundation Superstructure and Mat

2. The effective length of the girders should be the clear-span distance rather than the length between nodal positions. This can be accomplished by considering the part of the girder between the nodal point and the start of the clear span to be a rigid link (Figure 17). Similarly, the base of the columns should be rigid between the center and top surface of the base mat if the base of the columns is modeled at the midplane of the mat.

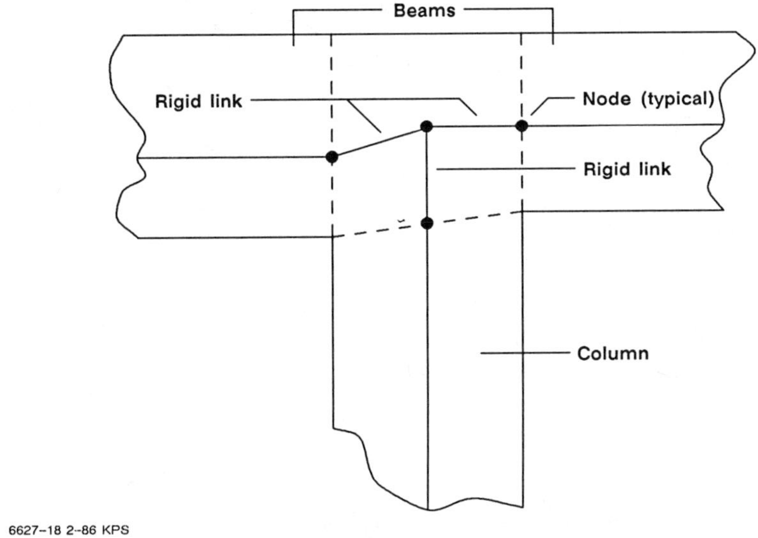

Figure 17. Modeling of a Member Connection

3. The loads that the machine applies to the foundation are transmitted to the deck through the soleplates. The soleplates are not generally at the centerline of the beams. Therefore, it is necessary to specify the loads in such a way that the torsion introduced into the system is taken into account. This may be accomplished by specifying a load acting at the girder centerline with an accompanying moment as shown in Figure 18. Alternatively, a node can be added at the point of application of load and rigid links specified between the node and the actual centerline of the girder, as shown in Figure 19.

4. For concrete foundations, the moment of inertia of the girders should be that of the uncracked concrete section because of the very low stress levels usually present in a turbine foundation. This assumption will yield results (forces, moments, and deflections) which are theoretically closer to the actual behavior of the foundation than if a cracked section were assumed. If the calculated stresses are higher than those compatible with an uncracked section, the analysis could be repeated using the effective moment of inertia as specified in the ACI Code (4).

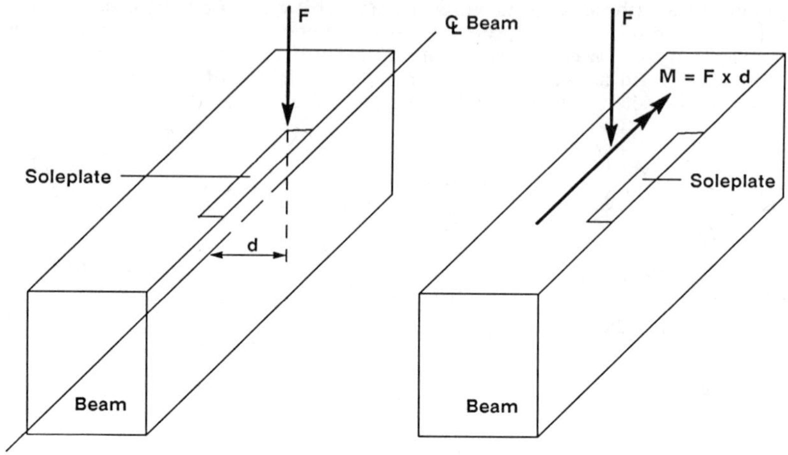

Figure 18. Modeling of a Load Applied To a Soleplate

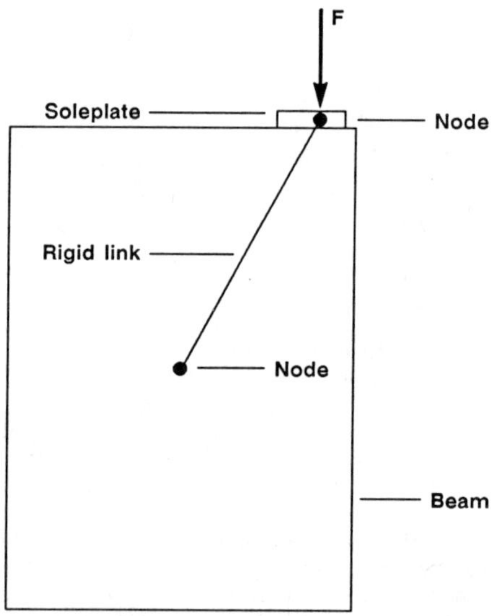

Figure 19. Modeling of a Load Applied To a Soleplate

5. It is recommended that nodes should be located in the deck model as follows:

   - At the junction of two or more beam members.

   - At intermediate points along the length of the beam where loads are to be applied, if the computer program used requires forces and moments to be applied at the nodes rather than at points along the span of the beam elements. An example of such points is the soleplate locations where the turbine and generator interface with the foundation.

   - At each generator and turbine bearing location.

   The results (displacements, forces, etc.) should be calculated at enough points along the length of the beams so that the shear and bending diagrams can be constructed. If the program used requires that a node be specified for each point where the results are to be calculated, it is desirable to specify nodes at regular intervals along the beam. If the program can calculate the results at points between nodes, the intermediate nodes between junction points are not necessary for a static analysis.

6. Enough elements should be used in the base mat model so that a fair estimate of the mat stiffness is obtained. A node should be placed at each column and an additional node between columns in both the transverse and longitudinal directions.

7. Concrete strength increases with time. As this occurs, the modulus of elasticity also increases. It is recommended that a value of E be used that represents the modulus at the time the machine goes into service. The modulus of concrete is given in Reference 4 as

$$E = 57{,}000\ (f'_c)^{1/2} \qquad [15{,}000\ (f'_c)^{1/2}] \qquad (5\text{-}1)$$

   where:

   E = modulus of elasticity of concrete in psi (kg/cm^2), and

   f'_c = compressive strength of concrete in psi (kg/cm^2).

Computer Programs and Results

Many computer programs are available that can be used for linear static analysis of a turbine foundation. Some commonly used programs are STRUDL (34), NASTRAN (25), ANSYS (8), SAP (30), and STARDYNE (33). The principal requirement for such a program is that it have the capability to analyze a three-dimensional space frame and that it have plate-bending elements to model the base mat and any shear walls in the foundation. In addition, the program selected should have a wide variety of loadings that can be applied to the foundation, and it should be able to include the special modeling considerations of the beams as described earlier.

Regardless of the program selected, the deflection of the nodal points in the foundation and base mat, the forces and moments acting on specified cross sections of each girder, and the force distribution in the plate-bending elements should be computed as part of the detailed static analysis.

Dynamic Analysis

The objective of a dynamic analysis is to determine if a turbine foundation will perform satisfactorily when subjected to dynamic loads, i.e., loads that vary with time. This section describes the recommended modeling procedures, the methods of analysis, and the procedures to maximize the dynamic response.

Models for Dynamic Analysis

The dynamic analysis of the turbine foundation requires a model of the superstructure that is composed of beam and plate elements similar to those required for the detailed static analysis. The turbine foundation mat, if modeled, is composed of plate-bending elements. The interaction between the mat and soil is modeled as a series of horizontal and vertical springs that connect the mat nodes to fixed supports. Recommendations on the soil spring modeling are presented later in this chapter.

The dynamic model should not be created independently of the static model, but should be created from the static model by incorporating the following:

1. Enough nodes should be specified along the length of the beams and columns so that all of the modes that have a natural frequency equal to or less than the operating speed of the machine will be calculated. The lower modes of the foundation, which involve deflection of the top deck as a unit, are not affected significantly by the number of nodes along the length of the beams. However, the higher modes, which involve differential deflection of the top deck, are affected by the distribution of the nodes. If not modeled with enough nodes, these modes may be entirely missed, leading to incorrect results.

   The recommended minimum number of nodes, n, to be placed along the length of a span is given by the larger of the following two values:

$$n \geq \frac{L}{\pi} \sqrt{\omega} \left(\frac{M}{EI}\right)^{1/4} \qquad (5\text{-}2a)$$

$$n \geq \left(\frac{1}{2} + \frac{L\omega}{\pi}\sqrt{\frac{M}{EA}}\right) \qquad (5\text{-}2b)$$

   where:

   E = the modulus of elasticity,

   M = the mass per unit length,

   L = the span,

   ω = the operating speed of the machine in radians per second,

A = the cross-sectional area of the member, and

I = the moment of inertia for weak axis bending of the member.

The above expressions are derived from the expressions for the bending and axial frequencies of a simply supported beam (37) and are based on the notion that if the nth natural frequency of the beam is at or below operating speed, at least n midspan nodes will be needed to calculate n modes using the discrete model.

2. The nodal masses can be computed either by the lumped-mass approach or by the consistent mass method.

   In the lumped-mass approach, the masses of the beam elements are lumped at the nodes at each end of the beam elements. This leads to a diagonal mass matrix for the structure. The consistent mass approach accounts for the distributed mass and the variation of deflection along the length of the beam. The mass matrix is nondiagonal, or full. For the same number of nodes, the use of the consistent mass matrix will yield natural frequencies which are theoretically more accurate than those computed from the lumped-mass approach (9), even though the difference may be minor for most practical problems.

3. For the dynamic analysis, Young's modulus can be computed using Equation 5-1. The compressive strength should represent the mean strength at the time the machine goes into service. This is different than that for static analysis, where it was recommended that the lower estimate of the compressive strength be used when the machine goes into service so that calculated deflections resulting from static loads would be equal to or larger than the actual deflections. This practice will not necessarily lead to a conservative calculation of the displacements resulting from dynamic loads, because a change in the modulus of elasticity changes the natural frequencies of the foundation, which in turn may cause an increase or decrease in the response of the structure.

   The value of the modulus under dynamic loads is slightly larger than that computed using Equation 5-1. Reference 6 presents the empirical relationship by which the dynamic modulus of elasticity can be calculated. However, Equation 5-1 is recommended because the structural stiffness is a function of EI, where E is the modulus of elasticity and I is the moment of inertia. The moment of inertia used is that of the uncracked section. As some cracking may occur, the use of an uncracked section moment of inertia overestimates the stiffness. The use of a slightly lower Young's modulus compensates for this.

4. It is recommended that 2% of critical damping for concrete elements and 1% of critical damping for steel elements be used for dynamic analysis for operating loads. The lower than normal damping values reflect the fact that turbine foundations are, in general, stressed to rather low levels under operating loads. A 7% damping for reinforced concrete elements may be used for emergency loads or when foundation elements are stressed to over 50% of their ultimate capacity.

5. The dynamic model of the turbine foundation should include the machine. Omission of the machine would introduce considerable error in the

calculated response of the foundation. There are two acceptable methods of modeling the machine on the foundation.

The first method of modeling the machine is to reduce it to a series of mass points lumped with the foundation model. These point masses are determined from the dead weight soleplate loading given by the turbine manufacturer. An example case, presented in Reference 11, illustrates this method of modeling. In this method, the dynamic interaction between the machine and the foundation is not taken into account, causing incorrect computation of the higher modes that involve relative displacement of the nodes on the top deck of the foundation. Therefore, the response of the foundation resulting from these modes could be erroneous.

The second and more accurate method of modeling the turbine-generator includes a dynamic model of the machine as well as the foundation (1, 7). This involves modeling the shaft as a lumped mass-beam system, modeling the bearing oil film as spring elements with specified stiffness and damping characteristics, and modeling the bearing pedestals as spring-mass systems. The stiffness and mass distribution of the turbine shell should also be incorporated if available from the turbine manufacturer. If this information is unavailable, the foundation engineer has no rational means of modeling the shell; in this event, the usual procedure is to lump the mass of the turbine shell with the foundation model based on the dead weight soleplate loading given by the manufacturer.

6. The effect of the foundation mat on the dynamic analysis of the superstructure depends on the type of dynamic load being considered. If the dynamic load is a high-frequency force such as a rotating unbalance, the effect of the base mat is small, that is, the base mat acts as a fixed base for the foundation. Therefore, the foundation may be adequately analyzed by using a model where the columns are assumed to be fixed at the mat (no translation and no rotation).

   On the other hand, if an exciting force such as an earthquake excites some of the foundation's lower modes, it is generally necessary to use a combined mat (crude model if necessary) and foundation model so that the mat flexibility and soil-structure interaction may be taken into account. A discussion of the soil-structure interaction effects is presented later in this chapter. A discussion of the effect of the base mat on the results of a dynamic analysis is given in References 12, 14, and 26.

Method of Analysis

Once the model has been created, it is necessary to perform a dynamic analysis to determine the displacement of the bearing locations and the axial forces, bending moments, shear forces, and torsional moments induced in the beams and columns of the foundation. These quantities are necessary to ensure that the foundation will meet the specified deflection and strength criteria.

The two methods that are generally used for dynamic analysis of turbine foundations are the modal analysis method and the complex frequency response method (21).

The modal analysis is appropriate when the damping and stiffness of the machine/foundation/soil system can be approximated as frequency independent.

If this is not the case, the natural frequencies and mode shapes must be computed for each forcing function frequency, leading to an uneconomical solution.

When the damping and stiffness cannot be approximated as frequency independent, the complex frequency response method is used. This method is general and is economical even when the damping and stiffness are not frequency dependent. A detailed description of the two methods is given in References 15 and 35.

Because of the specialized nature of the turbine foundation dynamic analysis, the following factors must be included in the analysis.

Range of Loading Frequencies - The dynamic response of the turbine foundation is of particular interest at the operating speed of the machine. However, it is also of interest for all speeds of the machine, from 20% to 120% of the operating speed, because the machine will operate at a range of speeds during startup and shutdown. Thus, it is desirable to construct a frequency response curve as shown in Figure 20 for design quantities, e.g., displacements, moments, etc.

Phasing of Unbalance Forces - A design value of the unbalance force arising from the rotor is specified by the turbine manufacturer at each bearing of the turbine-generator set, and the unbalance force acts simultaneously at more than one bearing. The unbalance forces should be arranged to yield the most probable maximum response, as illustrated in Figure 6.

In general, the arrangement of unbalance forces that causes maximum bearing displacement will not be the same as the arrangement that causes maximum bearing reactions. Also, the assumption that all the unbalance forces are in phase or aligned may not lead to a maximum displacement.

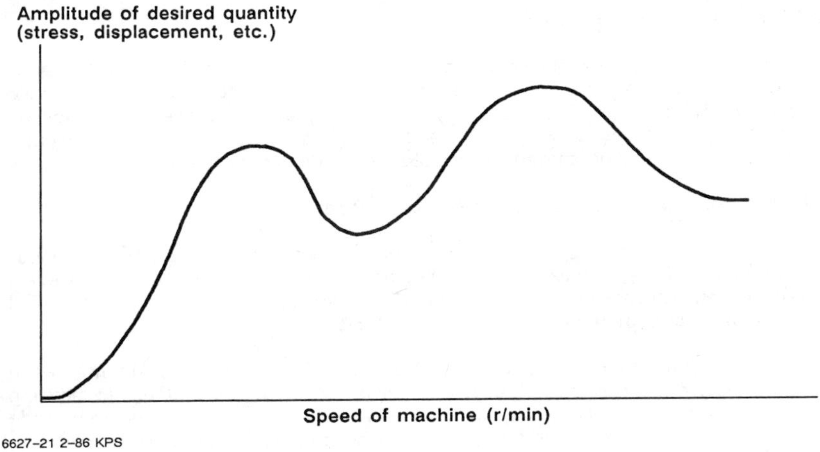

Figure 20. Response Spectrum for Loads Applied To Foundation

One method that can be used to compute the maximum displacements and forces is to analyze for the unbalance force at each bearing of the turbine-generator set for the frequency range of 20% to 120% of the machine's operating speed. An upper limit of the response for each frequency of the unbalance force is the sum of the values from each load acting separately. The most probable maximum response for each frequency of the unbalance force is the square root of the sum of the squares of the values for each loading acting separately. This method allows the engineer to create a plot similar to that shown in Figure 20 that will represent the maximum values of the desired response quantity as a function of force frequency.

An alternate procedure, which can be used in conjunction with the modal analysis, is presented in Reference 15. This method requires that several mode shapes of those modes near the operating speed be used to orient the unbalance forces for maximum displacement. However, this procedure cannot be employed when the complex impedance method is used because the mode shapes are not known.

Base Mat Analysis Procedures

A combined superstructure and mat model, in which the mat has enough nodes to adequately describe the stresses in the mat, yields results for both the foundation and the base mat design, making a separate analysis for the mat unnecessary. However, when the mat is not included in the turbine foundation model, or when only a crude mat model is included as part of the turbine foundation model, a separate analysis is necessary to compute the base mat design forces. For this analysis, the base mat is modeled by plate-bending elements and the soil or pile supports as spring elements.

The design moments and shears for the base mat are computed by imposing the maximum column reactions and by performing a static analysis. For this two-step analysis, the dynamic and static load reactions are applied as loads on the mat model. As stated earlier in this chapter, the two-step procedure is most commonly used.

Soil-Structure Interaction Effects

The presence of supporting soil affects the static and dynamic response of the turbine foundation. These effects are herein termed soil-structure interaction effects. In this section, the effect of soil-structure interaction on the static and dynamic response of turbine foundations is discussed and recommendations are presented for procedures to model these effects.

Effect of Soil-Structure Interaction on Static Response

Soil-structure interaction is central to the geotechnical aspects of foundation design, such as sizing the mat, checking total settlement, etc. However, soil-structure interaction for static loads is not important in the structural analysis and design of the deck and columns for the following reasons:

1. The machine alignment is affected only by differential displacements of the bearings, not by overall rigid body translation or rotation of the foundation deck. In conventional designs, the foundation members are sized to provide a rigid frame, thus minimizing the differential displacements resulting from soil-structure interaction effects.

2. Since the alignment is performed after the machine and foundation dead load have been active for some time, the majority of the soil-structure interaction effects due to machine and foundation dead loads occur prior to the initial alignment and thus do not affect the machine operation. Realignment performed periodically as part of routine maintenance accommodates any additional modest long-term soil-structure interaction.

3. The condenser vacuum loading and the thermal loadings are self-equilibrating within the foundation and thus their analysis is not affected by soil-structure interaction effects.

4. The foundation mat system for conventional designs is very rigid, thus minimizing differential displacements due to soil-structure interaction effects. In addition, the foundation is generally capable of accommodating moderate magnitudes of differential displacements by redistribution of member forces by inelastic action.

Effect of Soil-Structure Interaction on Dynamic Response

Soil-structure interaction is known to have a significant effect on the fundamental modes of turbine foundations (12). However, as the frequency of the exciting force increases, the dynamic response appears to become less sensitive to soil-structure interaction. Thus, the importance of soil-structure interaction for a given load condition depends on the extent to which the lower modes participate in the response.

The following paragraphs discuss the significance of soil-structure interaction effects for periodic and nonperiodic forces.

Periodic Forces - The principal periodic forces acting on a turbine foundation result from rotor imbalance. Whether associated with normal operating conditions or occurring as the result of an emergency (such as loss of blade), the forcing function is sinusoidal, having a frequency equal to the angular speed of the rotor and an amplitude proportional to the speed squared. For typical large turbine foundations, the fundamental frequencies are considerably lower than the operating speed (3, 12, 31, and 32), so the corresponding modes do not contribute significantly to the response. Consequently, turbine foundation vibration amplitudes at the operating speed typically exhibit little sensitivity to soil-structure interaction (3, 31, and 32).

During startup or shutdown, of course, the rotor speed must pass through the lower modes, so these modes will dominate the response at these frequencies. Thus, the low-frequency response is affected by the flexibility of the subgrade (17). For designing the foundation, however, the low-frequency response is of little importance because:

- the imbalance force amplitudes are much less than at the operating speed and are thus not critical for design; and

- the limits of vibration amplitude for acceptable performance (11) are greater than those at the operating speed.

Nonperiodic Forces - This category includes short-circuit torque and seismic loads. Because of the high frequency and short duration of short-circuit loads, the lower modes will not participate significantly in the response, so the effect of soil-structure interaction is negligible. On the other hand, the frequency content of typical seismic excitation is such that substantial, if not predominant, participation of the fundamental modes may be expected. Consequently, soil-structure interaction will significantly affect the seismic response of a turbine foundation.

Recommended Procedures

Based on the above considerations, it can be concluded that, except for seismic excitation, soil-structure interaction generally does not have a major impact on the design of the turbine foundation superstructure. Consequently, the following procedures are suggested for static and dynamic analysis.

Static Analysis - It is acceptable to model the deck and columns with the columns assumed fixed at the top of the mat. The mat is then analyzed separately, treating the fixed-end reactions from the columns as external loads. However, since the introduction of mat flexibility could result in some reduction of the member forces in the superstructure, the analyst may wish to use a combined model in which the mat is discretized with finite elements. In this case, a linear distribution of subgrade (or pile) reaction ordinarily will be sufficiently accurate to account for soil-structure interaction because of the rigidity of the mat. (The rigidity criteria in Reference 2 may be consulted to verify the validity of this approach.)

Dynamic Analysis - For a turbine foundation whose design is not governed by seismic loads or for which the response to harmonic excitation at frequencies well below the operating speed is not desired, a fixed-base model is adequate; otherwise, soil-structure interaction should be included. In the latter case, for reasonably homogeneous deposits, the soil may be modeled with lumped-parameter springs and dampers or zero-frequency impedance functions (modified, as necessary, to account for bedrock at shallow depth). The mat may be modeled as a rigid block or may be discretized with finite elements (the latter method provides a better representation of the mat, but difficulties arise in specifying equivalent distributed soil springs). For this reason, or when the soil deposit is highly nonuniform, the analyst may elect to model the soil using a more elaborate procedure, such as finite element or impedance matrix techniques.

Chapter 6

Deflections

Introduction

Turbine-generator foundation designers have used foundation design criteria that are based on the static deflections of the foundation under a specifically identified set of loading conditions. In the design of turbine foundations, deflections are first and foremost a functional requirement. The turbine-generator manufacturers have many reasons to limit the deflection of the turbine foundation to acceptable limits.

A typical shaft is supported by a number of bearings. Excessive differential deflections between adjacent bearings can induce large stresses in the shaft.

In addition to large shaft stresses, differential deflection between adjacent bearings may result in overloading some of the bearings while other bearings may be underloaded. Since each bearing is specifically designed for a range of minimum and maximum loads to optimize the bearing performance, turbine manufacturers recommend that the integrity of the bearings not be jeopardized by overloading or underloading any bearing beyond its acceptable margin of safety.

Finally, for all normal operating conditions, the minimum running clearances between the stationary and rotating elements of the machine must be maintained in the turbine interstage and shaft gland steam seals and in the generator hydrogen seals. One way to achieve smooth operation of a machine is to limit the total deflection, and in turn the differential deflection, of various structural components of the turbine foundation.

In order to achieve this desired objective, the turbine manufacturers have used various methods of specifying the deflection limits of turbine foundations.

History of Deflection Criteria

Originally, when the units produced less than 10,000 kW, turbine-generator foundations were designed to standards established by the foundation designers. These early foundations were designed conservatively, since no deflection limits were given by the turbine-generator manufacturers. The loading, usually turbine-generator weight and condenser loads, was provided by the equipment suppliers.

A number of these foundations, each of which had a proven history of satisfactory turbine-generator operation, were reviewed. Deflections at selected points were calculated and compiled. From this data some general conclusions about the deflections of the turbine-generator foundations were reached; that is, experience indicated that foundations designed to within these "established" deflection limits would result in satisfactory performance.

However, as the machines became larger, it became extremely difficult for the foundation designer to meet these deflection criteria. A brief review of various methods of specifying deflection limits with the advantages and disadvantages of each follows.

Absolute Deflection Criteria

Past experience has shown that limiting the absolute deflection of various foundation components to within acceptable limits will ensure that the turbine-generator will run smoothly. An example of an absolute deflection criterion is that the total allowable vertical elastic shortening of the columns under a given loading is limited to 0.020 inch (0.5 mm).

The advantages of using this type of criteria are that the resulting design will be both safe and conservative and the turbine-generator units will operate satisfactorily. Also, the foundation designer can complete the design without resorting to numerous interaction calculations that require input from the turbine-generator manufacturer. The main disadvantage is that the resulting design requires massive piers, which eliminates valuable room under the deck for piping and equipment and increases the cost of the foundation. In addition, the rationale for the governing loading is not always clear or easy to follow.

Radius of Curvature Criteria

The concept of defining the deflection tolerances by a shaft radius of curvature has been used in the past by some turbine manufacturers and foundation designers. The radius of curvature is defined as the radius of the circle determined by any three points that lie in a plane in space. This criterion generally requires that radius of curvature be checked at each set of three adjacent transverse girders for both the vertical and transverse directions. The acceptable limits of vertical and transverse radii that describe the shaft curvature caused by loadings (except dynamic loading) subsequent to initial alignment vary from one machine to another; however, one general criterion that has been widely used is 75 miles (120 km) for an 1800-r/min machine and 150 miles (240 km) for a 3600-r/min machine.

The advantage of this criterion is that it is generally easier to meet than the absolute deflection criterion, since the radius of curvature criterion considers the differential deflection between adjacent beams or bearing supports rather than a total deflection. The disadvantage of this concept is that an acceptable radius of curvature value in a range of 75 to 150 miles (120 to 240 km) results in a conservative design that forces the foundation designer to provide a stiff deck similar to one designed by the absolute deflection criterion. As a consequence, the criterion is difficult to meet for large foundations.

Influence Coefficient Approach

In order to more accurately determine the effects of displacement of the foundation under operating loads, several turbine manufacturers have developed deflection criteria that utilize the influence coefficient approach. These criteria define the amount of misalignment that a turbine-generator rotor system can withstand without imposing unacceptable loads on the rotor bearings or unacceptable bending stresses in the rotor system.

The foundation designer can calculate the deflections that may result from various loads that are applied to the foundation after the initial turbine alignment. He can then use the influence coefficients provided by the turbine-generator manufacturer for that specific unit to determine whether the foundation response is likely to adversely affect the turbine-generator operation or reliability.

An influence coefficient as used here is the factor by which a deflection is multiplied in order to obtain the response at a specific location. The response may be the load at a bearing support point, the bending stress in the shaft, or a normalized general force. These responses are compared against a given set of allowables (minimum and/or maximum bearing load, maximum stress, etc.) to ensure that the deflection of the foundation does not adversely affect the operation and reliability of the machine.

As an example, consider the single-span, propped cantilever beam shown in Figure 21. If the beam is straight and has no initial support deflection, the response at each of the locations shown in Figure 21 is R_1', R_2', and R_3'. These responses could be beam member forces, support reactions, beam deflections, or stresses. Due to support settlement or an initial temperature deflection, the three points will displace vertically upward (+y) or downward (-y). These deflections are labeled y_1, y_2, and y_3. These responses of the beam system are a function of the beam geometry, support stiffness, and the distances between points of interest. The particular response due to a displacement can be expressed in terms of a series of influence coefficients (C), which are unique for a specific beam and a particular response. For example, for a given beam, there is a unique set of three coefficients to describe the bending stress at Location 2 due to a deflection at each of the three locations. In general, this can be expressed for Location 2 as:

$$R_2 = R_2' + C_1y_1 + C_2y_2 + C_3y_3 \qquad (6\text{-}1)$$

where:

R_2' = the initial response at Location 2,

C_i = the change in response at Location 2 per unit of vertical deflection of Location i, and

y_i = the vertical deflection of Location i from the initial position.

Note that this response may be deflection, beam bending or shearing stress, support reactions, beam member forces (moment, shear), etc., depending on the quantity of interest. Note also that the position of the point of interest as well as the required response determine the sign of the quantity, C_i.

A specific example of the influence coefficient approach for turbine-generator foundation design follows. Figure 22 shows a simplified line drawing of the first two casings of a turbine-generator with a shaft system supported by

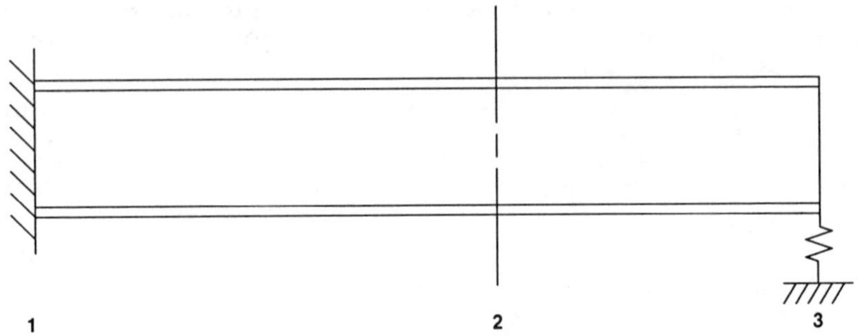

Figure 21. Propped Cantilever Beam

four bearings. The rotors are precisely aligned with one another in the cold condition, and the couplings are bolted together. The loadings on the four bearings in Figure 22 would be P_1', P_2', P_3', and P_4', provided the machine could run with undisturbed shaft alignment. During normal operation, however, all the bearings are displaced vertically upward (+y) or downward (-y) due to foundation deflections. The various amounts by which the bearings are loaded and unloaded as a result of such shaft misalignment depend upon the design of the bearings, the configuration of the shaft system, and the distances between the bearings. The influence coefficients for this example give the amount by which the loading on a bearing changes per inch (mm) of bearing support displacement. For example, the load on Bearing 2 as a result of shaft misalignment is:

$$P_2 = P_2' + (-C_1)y_1 + (+C_2)y_2 + (-C_3)y_3 + (+C_4)y_4 \quad (6\text{-}2)$$

where:

P_2' = calculated loading of Bearing 2 with correctly aligned shaft system in psi (kg/cm^2),

C_i = change in loading on Bearing 2 per inch of vertical deflection of Bearing i in psi/in. ($kg/cm^2/mm$), and

y_i = vertical deflection of Bearing i from the position of initial correct shaft alignment in inches (mm).

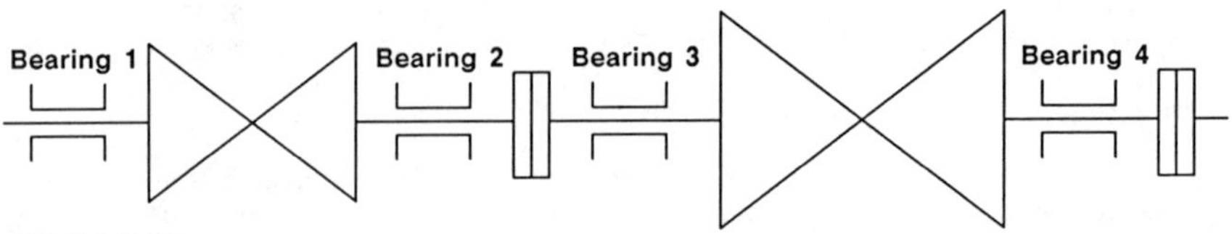

Figure 22. Four-Bearing Turbine Shaft System

In this particular example, the influence factor C_2 will be positive and have the largest value, reflecting the fact that an upward displacement of Bearing 2 strongly increases the load on the same bearing, whereas downward displacement results in a correspondingly large decrease in load. The negative factors C_1 and C_3 denote that upward displacement of Bearings 1 and 3 tend to unload Bearing 2. The value of C_3 will, of course, be considerably larger than C_1 due to its proximity to Bearing 2. The effect of other bearing displacements on the loading of Bearing 2 becomes smaller with increasing distance from Bearing 2 so that the factor C_4 will have the smallest value.

In general, the vertical and horizontal coefficients at each of the bearings are required from the turbine-generator manufacturer in order to use this approach. After the results of the influence coefficient analysis are obtained, the responses are compared to the allowable values given by the manufacturer (e.g., maximum and minimum loads for bearing loading; maximum stress, either vertical or horizontal direction or a combination of the two; or a normalized foundation rigidity value). The units as well as the locations of the required displacements are specified by the turbine-generator manufacturer.

Some forms of misalignment of certain shafts can have an adverse effect on the bearing loadings but need not greatly increase the bending stresses in the shaft systems. It is equally possible for some shaft systems to be vulnerable at particular locations to high bending stresses with certain forms of shaft misalignment that do not result in any substantial alteration of the loads on the individual shaft bearings. When the combined effect of the foundation deflections is found to exceed the allowable response at one particular point, it should be recognized that a design change at another location can be a feasible solution.

Some turbine-generator manufacturers provide one set of influence coefficients to determine the effects of foundation displacement on the bearing loads and a different set of coefficients to investigate the effects on the shaft bending stresses. Other manufacturers combine the effects of bearing loads and shaft stresses into a single set of coefficients. In either case, the analysis by the foundation designer is the same.

Load Cases for Calculation of Static Deflection

Any deflection calculation that presumably intends to safeguard the shaft alignment should consider only those loads that are applied to the structure after the alignment is completed. Since the method of calculating the deflection is based on a static analysis approach, the loadings included in the deflection calculation should include all post-alignment static normal operating loads. The calculation should include the effects of the following post-alignment loadings:

- condenser vacuum load,
- normal torque load,
- effect of thermal loading due to machine expansion-contraction, and
- piping loading.

The following should be considered if significant:

- effect of foundation settlement subsequent to the alignment; and
- effect of changes in the temperature surrounding the foundation, e.g., effect of sun shining on one side of an outdoor unit.

Experience indicates that the condenser vacuum load and normal torque load can cause appreciable foundation deflection subsequent to the initial alignment. The effects of thermal gradients within the foundation due to heat transfer from the equipment and the effects of creep may need to be considered for low-tuned foundations.

Dynamic Response Considerations

Historically, the dynamic coupling between the turbine-generator and the foundation on domestic units has not been a serious consideration for the foundation designer. The traditionally massive and rigid conventional concrete foundations have resulted in little interaction between the foundation and the power-generating equipment. Consequently, dynamic response criteria were not required and the turbine designer and the foundation designer could work without regard to the dynamic characteristics of the other's design.

Since turbine-generators have grown in rating and size and a trend toward more flexible foundations has developed, increasing attention has been given to the dynamic characteristics of the turbine/generator/foundation system. The concerns of the turbine manufacturer are related to the reliability of the machine and the impact of the unit on the local foundation deck environment. Small amounts of inherent unbalance in the turbine-generator can be amplified through resonance with the foundation to produce excessively large bearing forces. Large relative bearing displacements could also produce large bearing forces and high rotor bending stresses. These adverse conditions could affect the availability of the unit or lead to a forced outage. In addition, excessive foundation deck response could be judged uncomfortable from a "human sensitivity" consideration, leading to dissatisfied clients.

In order to ensure satisfactory and reliable operation of the turbine-generator and to provide a working environment around the unit that is judged acceptable, turbine manufacturers have started specifying dynamic response characteristics. These characteristics have taken the form of a maximum deck response under normal and emergency conditions or maximum levels of other vibratory characteristics.

One vibratory characteristic that has been suggested by the International Organization for Standardization (ISO) for describing the vibratory state of a machine is known as the "vibration severity." The vibration severity is defined as the root-mean-square value of the vibration velocity. In ISO publications 2372 and 3945 (19, 20), the philosophy behind the use of the vibration severity factor is described along with recommended limits for the factor. The limits are established for both "rigid" and "flexible" foundations. A flexible (rigid) support is defined as one where the fundamental vertical frequency of the machine and support system is lower (higher) than its main excitation frequency. Measurement points are at or near each bearing on the bearing housing.

The ISO qualifies the recommended limits by stating that the "severity values are useful for comparison purposes or general machine acceptance, but they are not intended to reflect specific design criteria and/or specifications which may exist for a given machine. Thus, in some instances, requirements either more or less stringent, or additional to the recommended limits. . ., may be invoked." Therefore, the foundation designer should consult with the turbine manufacturer to determine the specific requirements for a given project.

Chapter 7

Design

This chapter presents the design criteria and procedures commonly used for the design of large concrete turbine-generator foundations.

Special design problems and details associated with the design of concrete turbine-generator foundations are also discussed.

Acceptance Criteria

Turbine-Generator Manufacturer's Criteria

Major turbine-generator suppliers generally specify their performance criteria for the turbine-generator foundations, with particular regard to foundation deflection and machine alignment, and often state their preference for the type of support system (i.e., concrete, steel, composite, or spring-supported concrete/steel).

Codes

In addition to the manufacturer's criteria, applicable sections of the following codes and standards are generally used in the design of turbine-generator foundations:

- American Concrete Institute (ACI): "Building Code Requirements for Reinforced Concrete" - ACI 318.
- American Institute of Steel Construction (AISC): "Specification for the Design, Fabrication and Erection of Structural Steel for Buildings."
- Other codes, standards, and regulations as applicable.

Architect-Engineer Criteria and Standard Practice

Each particular foundation has specific functional requirements, and strict adherence to any of the codes and standards mentioned previously may be inappropriate. Special criteria and code deviations have been established, where required, based on rational analysis, research, and experience.

For reinforced concrete foundations, the following design criteria, in addition to the applicable codes, are commonly used:

- The minimum dimensions of beams and columns are usually controlled by stiffness rather than strength requirements and are discussed in Chapter 4.

- The compressive strength of concrete (f'_c) most commonly used for the foundations is 4,000 psi (27.5 MPa). The yield strength of reinforcement is 60,000 psi (410 MPa).

- The minimum reinforcement is listed in the following subsections.

Columns -	Longitudinal reinforcement:	0.5% of gross concrete area.
	Ties:	No. 5 bars with a maximum spacing of 18 inches (450 mm).

All longitudinal bars are laterally supported by ties unless they are 12 inches (300 mm) or closer along the tie to the next laterally supported bar. Lateral support is provided by the corner of a tie with an included angle of not more than 135°.

Section 10.9.1 of ACI 318-83 requires the minimum longitudinal column reinforcement to be 1% of the gross concrete area. However, Section 10.8.4 allows the use of a reduction in effective area of up to 50% for the calculation of minimum reinforcement. Therefore, the minimum reinforcement may be as low as 0.5% of the gross area. If the reduced effective area is used for strength calculations and minimum reinforcement, the full gross area should still be used for the column stiffness calculations.

Beams -	Top and bottom:	0.25% (each) of gross concrete area.
	Sides:	0.1% (each) of gross concrete area.
	Stirrups:	No. 5 closed stirrups minimum, with maximum spacing of one-fifth of beam depth but not to exceed 18 inches (450 mm).
Mat -	Top and bottom:	0.1% (each) of gross concrete area each way.

General Requirements - (a) All faces of concrete members are reinforced with bars spaced not more than 12 inches (300 mm) on centers. (b) In view of the large bar diameters generally employed, careful attention must be given to anchorage and development details. (c) Splices should be located in compression zones, where possible. (d) A maximum bar size of No. 11 is preferable, although in some cases No. 14 or No. 18 bars in conjunction with mechanical splicing may be used to advantage. (e) The clear distance between parallel bars in a layer should be 6 inches (150 mm) or more, unless lateral reinforcement sufficient to prevent bond splitting is provided.

Design Procedures

Load Combinations

The foundations are designed to withstand the following load conditions:

- Operating - Loads expected during normal plant operation.
- Accident - Dynamic loads expected to occur very rarely during the plant lifetime. These loads include: generator accident (short circuit or synchronization out-of-phase), turbine accident (bowed rotor or missing last-stage bucket), and earthquake.

The following individual loads are considered in various possible combinations.

- DF - Foundation Dead Load
- DM - Machine Dead Load
- DE - Dead Load from Other Equipment and Piping
- LL - Live Load
- CD - Condenser Dead Load
- CV - Condenser Vacuum
- QN - Normal Torque (Turbine and Generator)
- TM - Machine Thermal Expansion/Contraction
- TF - Load Due To Thermal Gradient in Foundation
- TP - Piping and Valve Reaction (Excluding Dead Load)
- NB - Normal Machine Unbalance Load
- EE - Seismic Load
- QE - Generator Emergency Torque
- AB - Load Due To Bowed Rotor
- AM - Load Due To Missing Rotor Blade

For the precise definition and explanation of these loads, see Chapter 3.

Load Factors

Load factors for the ultimate strength design of concrete members are included in the following equations. The factors are based on the maximum anticipated magnitude and duration of the loads, and the probability of their simultaneous occurrence. These are general recommendations and factors other than those recommended here may be considered where applicable.

Operating Condition -

$$U = 1.4 (D + C) + 1.7 (LL + QN + TF + TM + TP + NB) \quad (7\text{-}1)$$

where:

$D = DF + DM + DE$, and

$C = CD + CV$.

Accident Conditions -

Generator Emergency:

$$U = (D + C + LL + QE + TF + TM + TP + NB) \quad (7\text{-}2)$$

Bowed Rotor:

$$U = (D + C + AB + LL + QN + TF + TM + TP) \quad (7\text{-}3)$$

Missing Rotor Blade:

$$U = (D + C + AM + LL + QN + TF + TM + TP) \quad (7\text{-}4)$$

Seismic:

$$U = 0.75 (1.4 [D + C] + 1.7 [LL + QN + TF + TM + TP + NB + 1.1EE]) \quad (7\text{-}5)$$

It should be noted that some of the individual loads included in the preceding load combinations may not be applicable nor significant in many instances. They may also have different values at different times during plant testing, startup, operation, or maintenance. These different conditions should be considered for each applicable load combination.

Special Considerations

Deep Beams

A number of turbine-generator foundation members may qualify as "deep flexural members" by the criteria of ACI 318-83. Such members may be designed in accordance with Section 10.7 of ACI 318-83 or, alternately, they may be designed as ordinary flexural members, in which case additional longitudinal reinforcement is provided in accordance with Section 10.6.7. Use of the former procedure may result in saving some reinforcing steel, but for deep beams of the proportions found in turbine-generator foundations, the savings will be small.

Shear and Torsion

Some of the flexural members of a turbine foundation are subjected to substantial shear and torsion forces. Examples are the longitudinal beams supporting the generator stator and the low-pressure turbine exhaust hoods. These members are usually short and heavily loaded, principally along one edge, resulting in relatively high torsion forces (see Figure 19). Combined shear and torsion strength of these members should be evaluated by the procedures of Chapter 11, Sections 11.3 and 11.6, of ACI 318-83, taking into account the effects on strength of simultaneous bending moments and axial forces at the section under investigation.

For beams loaded along their top surface, design sections for shear and torsion are taken at a distance "d" from the face of the support unless the shear at the face of the support is radically different, as in the case where a heavy load or beam reaction is applied within this distance. In this case, the design section should be taken at the face of the support. The distance "d" is defined as the dimension from the compression face of the concrete to the centroid of the tension reinforcement. Shear reinforcement and torsional reinforcement must meet the requirements of Sections 11.5 and 11.6 of ACI 318-83, respectively, and should be in the form of closed stirrups.

In special cases where loads are applied within the lower two-thirds depth of the beam, lateral reinforcement should be provided to transfer the load to the top of the beam.

Anchor Bolt Design

Anchor bolts for turbine-generator components are specified by the manufacturer with regard to number, size, location, elevation, threaded length, and preload. When required, the sleeve diameter, length, and elevation are also specified. The embedment of the bolt and its details are the responsibility of the foundation designer.

Anchor bolt data provided by the turbine-generator manufacturer assume that clearance is provided between the bolt and the soleplate, such that the bolt has only vertical loading; i.e., no shear force is exerted on the bolt. Shear forces are transferred to the foundation by the soleplate-grout interface and other anchorage devices.

To ensure a ductile failure of the anchorage, the anchor bolt embedment should be designed for a force at least equal to 1.25 times the ultimate strength of the bolt.

A ductile anchor bolt system may not be possible when bolt preload is specified by the turbine-generator manufacturer. Anchor bolts with preload may be sized to limit the stress below the yield stress, resulting in an oversized anchor bolt for which the embedment cannot resist 1.25 times the ultimate strength. If this situation arises, the embedment should be designed for the largest resisting force available but not less than the preload multiplied by 1.25 times the applicable load factor. This design approach should be used only when absolutely necessary.

Anchor bolts may be of the embedded type or the through type. A through-type bolt is advantageous for field adjustment, but location and clearance of the bolt ends should be verified by the turbine-generator manufacturer.

An embedded bolt may have a sleeve to permit adjustment. If a sleeve is provided, the inside sleeve diameter should be 1 inch (25 mm) larger than the bolt. In all cases the sleeve length should permit the bolt to move by double curvature. For bolts larger than 1-1/2 inches (38 mm) in diameter, the sleeve may not be effective. An alternate method for permitting bolt movement is shown in Figure 23. This arrangement permits movement without any deformation of the bolt.

All bolts, whether or not sleeves are used, should be set in place by a template, which is fabricated such that each bolt is tied together by the framework.

Anchor bolt design should be in conformance with ACI 349, Appendix B (5).

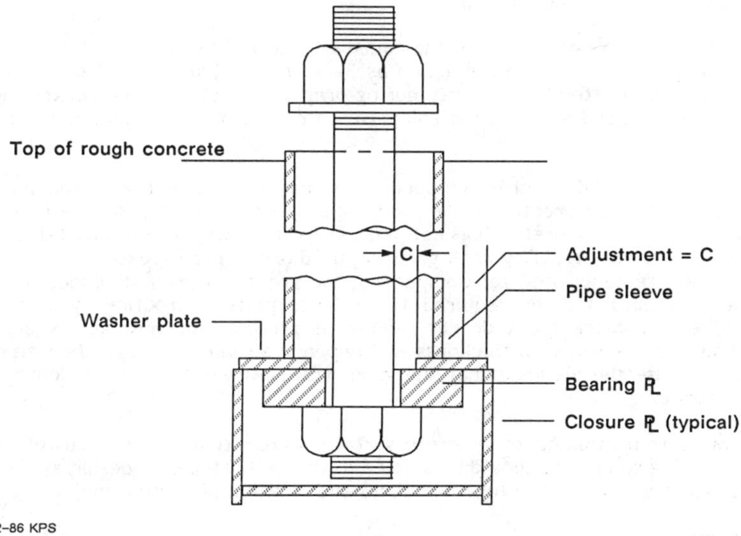

Figure 23. Adjustable Anchor Bolt

Design Details

Soleplates

Turbine-generator soleplates are large steel plates that are grouted and bolted to the top of the foundation and that support the turbine-generator components. The soleplates, furnished by the turbine-generator manufacturer, serve to distribute the loads and forces from the machine to the foundation. They must be positioned accurately and must give solid, unyielding support in order to set and hold the unit in proper alignment.

Normal loads are transmitted in compression directly through the grout to the concrete foundation below. Lateral loads are transmitted by bond or friction, or by grout plugs threaded into the underside of a soleplate and projecting into the grout. The method used depends on the magnitude of the lateral load. When used, grout plugs are furnished by the turbine-generator manufacturer.

The surface of the concrete beneath the soleplates must be sound and strong to ensure adequate support. It should be treated the same as a horizontal construction joint in a heavily loaded column; it should be properly cured, thoroughly cleaned to remove all laitance, and roughened to expose fresh aggregate faces. The grout between a soleplate and the concrete foundation must also be sound and strong.

The thickness of the space that must be allowed for grout will vary from as little as 1 inch (25 mm) to as much as 3-1/2 inches (90 mm), depending on the amount of concrete chipped away during preparation of the surface, the method used in setting and grouting the soleplates, the size of the plate, and the type of grout used.

The amount of concrete chipped away may depend on the customary practice of the turbine erector or the grouting method employed, irrespective of the soundness of the concrete. It is not unusual to chip away as much as 1-1/2 inches (38 mm) of concrete. Dry pack grouting methods require greater thicknesses to accommodate tools used for compacting the grout. A grout thickness of 3-1/2 inches (90 mm) may be required for a large plate. Grouting methods using flowable non-shrink grout do not require as great a thickness as the dry pack method, but the required thickness will depend to some extent on the size of the plate because the distance that the grout will flow varies with the depth of the grout space.

Matters pertaining to concrete surface preparation and methods of setting and grouting soleplates should be settled in the early stages of design as they will influence the size of structural members and location of reinforcing.

Grouting

The transfer of loads from the turbine-generator soleplates to the concrete foundation is accomplished by placing grout between the two surfaces. Although the strength must be adequate, the dimensional stability of the grout is the most important consideration. To have full contact with the soleplate, the grout must be free of voids and must not be subject to settlement by drying shrinkage. Portland cement-sand dry pack and several premixed manufactured grouts meet the above requirements.

The selection of a grout type is usually determined by the engineer and/or the turbine-generator manufacturer. Once a grout type is selected, the mixing, placement, and curing should be per the manufacturer's instructions. For additional information on grouting, the reader is referred to the ACI Committee 351 Report entitled, "Grouting for Support of Equipment and Machinery."

Reinforcement

Metal reinforcement must conform to ACI Standard 318, "Building Code Requirements for Reinforced Concrete." In areas of high seismic risk, the superior ductility of ASTM A706 reinforcement should be considered when selecting reinforcement. Fusion welding of bars, other than ASTM A706, is prohibited.

Reinforcing bars should be designed so that the bars can be fabricated and installed in their intended location without interference with embedments, conduits, or other bars and without undue congestion. Special attention should be paid at splice locations and intersections of members where bar congestion is most likely to occur. It is recommended that large-scale sketches be prepared at locations of reinforcing bar congestion to ensure that there is sufficient space for proper bar placement and concrete flow. Bar bends should be specified to have the minimum diameter permitted by the code.

Jacking Posts

Jacking posts are temporary fixtures installed on the operating deck to support horizontal jacks used to shift major components into alignment during erection of the turbine-generator. Typically, jacking posts are located near the end of each side of major components. The number, location, and projection of jacking posts should be specified by the turbine-generator manufacturer. Where this information is not specified and the designer determines that the posts are required, the posts normally project 18 inches (150 mm) and are located at each major component. The post loading is considered to be equal to the component weight at installation, multiplied by the coefficient of friction between steel and concrete, which can be conservatively taken as 0.5.

Jacking posts are usually lengths of steel rail or rolled shapes cast in the operating deck; however, a structural steel bracket or rolled section bolted to the deck or a similar fixture could also be used.

Design of the jacking posts above the rough concrete should be in accordance with the AISC "Specification for the Design, Fabrication and Erection of Structural Steel for Buildings." Concrete design should be in accordance with ACI 318, with the jacking load having a load factor of 1.7 and the bearing width equal to the flange width. Flange bending of the embedded section of the post can be neglected. The concrete bearing stress distribution is assumed linear. Based on standard practice, the minimum embedment length is usually 2 feet (0.6 meter). Allowable stresses may be increased by one-third due to the temporary nature of the load.

To facilitate the placing of reinforcing, the embedded jacking posts should be no larger than a W12. Reinforcing should be detailed to avoid interference with the post. If holes in the posts are necessary, the affected area should be accounted for in the design.

When erection of the turbine-generator has been completed, jacking posts or fixture anchor bolts should be flame-cut flush with the rough concrete surface prior to the placing of the finish topping.

Reference Points

Permanent elevation reference points set in the turbine-generator foundation deck are useful for monitoring differential settlement or differential thermal expansion. A typical reference point is a stainless steel ball welded atop a concrete anchor, cast monolithically in the structural concrete or set in non-shrink grout, with the ball recessed below the finished floor. A covered access port set flush with the floor protects the reference point from accidental damage or undue wear by traffic. In plan, a point is located near the center of each foundation column.

Monitoring of relative or differential elevations will suffice in most cases. One of the reference points is designated the benchmark and all readings are referred to that point. If measurement of absolute settlement is required, it can be facilitated by a benchmark consisting of an Invar rod in a steel casing grouted into the subgrade below the seat of consolidation. The rod and casing should be extended upward through the foundation to a point just below the surface of the finish floor.

It is essential that initial readings be made soon after the final alignment of the unit before subsequent settlement occurs. Readings must be accurate to within 1/1000 inch (0.025 mm). Such accuracy requires careful work by a skillful surveyor using a precision vernier theodolite. To ensure accuracy, a second set of readings should be made.

Because of the infrequent need of settlement data and the uncertainties inherent in the accuracy of the data acquired from reference point readings, not all engineers are persuaded that reference points need to be routinely installed. Some prefer to use an indirect measurement such as the change in coupling alignment when the need arises.

Concrete Topping

Floor areas atop the turbine-generator foundation usually receive a separate finish to match the surrounding operating room floor. The separate finish may be quarry tile or a topping of concrete with a smooth trowel finish.

The topping conceals the rough surface of the structural concrete, giving the floor a smooth-finish appearance. It should be installed after completion of the turbine-generator erection to avoid the possibility of damage from erection operations. Thickness of the topping may be as little as 3 inches (75 mm) or as much as 6 inches (150 mm). The choice of thickness is influenced by a number of interrelated factors: (1) required minimum concrete cover over reinforcement in the structural concrete, (2) accommodation of anchors for appearance lagging, (3) type of finished floor, and (4) soleplate elevations and thicknesses. The choice should be made in the early stages of design work, as it will set the location of the top of the structural members. Thick topping should be reinforced with welded wire fabric.

Demarcation of the edges of the topping along the expansion joint around the foundation may be achieved with structural plates or angles anchored to the structural concrete. These will protect the edges of the topping from chipping and control the width of the expansion joint around the foundation.

Concrete topping materials and installation should conform to good practice such as that described in "Concrete Floors" published by the Portland Cement Association (27).

Anchor Blocks and Centering Guides

The transverse alignment of adjacent turbine-generator components, which is required to prevent possible bearing and/or rotor failure, is maintained by the use of centering guides (also known as anchor blocks) attached to the foundation. Figure 24 shows typical locations for centering guides. Similar devices may also be required for longitudinal restraint.

The location, details, and loading for the guides are given by the turbine-generator manufacturer. The type of guide varies with the requirements of the manufacturer.

To maintain the close alignment required for proper operation, the guide should be extremely rigid. The deflection relative to the transverse beam is limited to no more than 0.005 inch (0.125 mm), unless different requirements are stated by the manufacturer.

Design of the embedment should be in accordance with ACI 349, Appendix B (5). Design of the steel components should be in accordance with the AISC Specification.

Restraints and Hangers

The turbine-generator foundation is designed to support and restrain the piping that is routed within the perimeter of the foundation. The largest concentration of piping is located between the high-pressure and low-pressure turbines.

Provisions are made to support the piping by use of embedded plates, angles, etc., or cast-in-place inserts. If expansion anchors are used, care should be taken not to drill through the reinforcing steel. Two approaches are available for providing support of the hangers. The first is to supply the necessary embedments for the hangers at the support locations. To do this, the exact

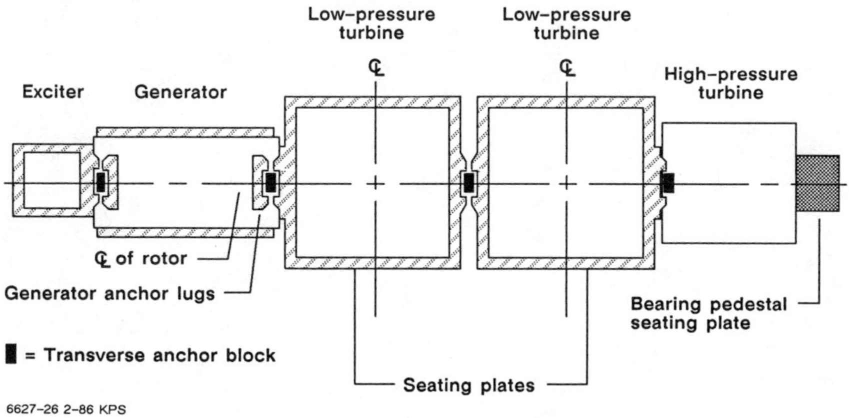

Figure 24. Typical Location of Anchor Blocks

location and load for the hanger support must be known. The applicability of this method is dependent upon the project schedule. The second method is to provide blanket coverage for support of the hangers when their location is not known. This could be accomplished by simply providing embedments at the corners of beams and columns or by providing a gridwork of embedments. Selecting the approach to use is left to the judgment of the designer. If possible, large, high-pressure piping should be supported at predetermined locations.

Hanger loads vary with each pipe classification and system. Typical ranges of hanger loads are from 10 to 40 kips (45 to 180 kN) for large, high-pressure piping, and from 1 to 5 kips (5 to 25 kN) for smaller piping. Design of the embedments should be in accordance with ACI 349, Appendix B (5).

Pipe Sleeves

Pipe sleeves are placed in the operating deck and columns as required by the manufacturer of the turbine-generator and the mechanical and electrical requirements of the architect-engineer. Where practical, sleeves are placed in areas of low shear and penetrate the center of the member to minimize disruption of the reinforcing. The diameter of the pipe sleeve is determined by the size of the pipe, insulation, lagging, movement, and clearance.

Where reinforcing is interrupted, details equivalent to an opening of the sleeve size should be used.

Chapter 8

Steel Foundations

Introduction

The first steel foundation was built in Germany in 1936, steel having been selected due to poor subsoil characteristics. The increased flexibility of this design reduced the vibratory forces transmitted to the subsoil, thereby reducing the probability of differential settlements and excitation of adjacent structures.

Since 1936 more than 1200 steel foundations have been designed and constructed in Europe. The largest steel foundation in Europe is supporting a unit with a capacity of 660 MW. At least 15 steel foundations have been constructed in North America. Thirteen of these were erected in Canada since the 1960s and are in the 150- to 400-MW range, and two are located at a unit in the central United States and support machines of a 500-MW rating.

Structural Design

Foundation Deck

The foundation deck (Figure 25) consists basically of a frame structure supported by slender columns and provided with cutouts for the high-, intermediate-, and low-pressure sections of the turbine, and for the generator. Additional openings for valves, steam inlet chests, air ducts, oil piping, etc., interrupt the superstructure of the frame. Pedestals for supporting the journals and intermediate structures to equalize any differences in height are welded to the deck, or the deck is built to suit the elevations of individual components.

The machine alignment demands axial bending resistance of the longitudinal beams. These must also have sufficient torsional strength to transfer the operating torque in the axial direction. Furthermore, the deck has to be rigid enough to maintain the alignment of the machine even if substantial unbalances from machine defects occur, and to sustain other impacts as well. The alignment accuracy requires a sufficiently rigid deck, which is also important for the vibrational characteristics. The deck is evolving more and more as a base frame of the machine and is therefore becoming a part of the machine itself.

To obtain the necessary bending and torsional rigidity, box girder construction with additional separate stiffeners is provided. The arrangement of stiffeners separating the structure into cells, Figure 26, should ensure that no individual plate or wall is in resonance with the frequency of the operating speed or its harmonics.

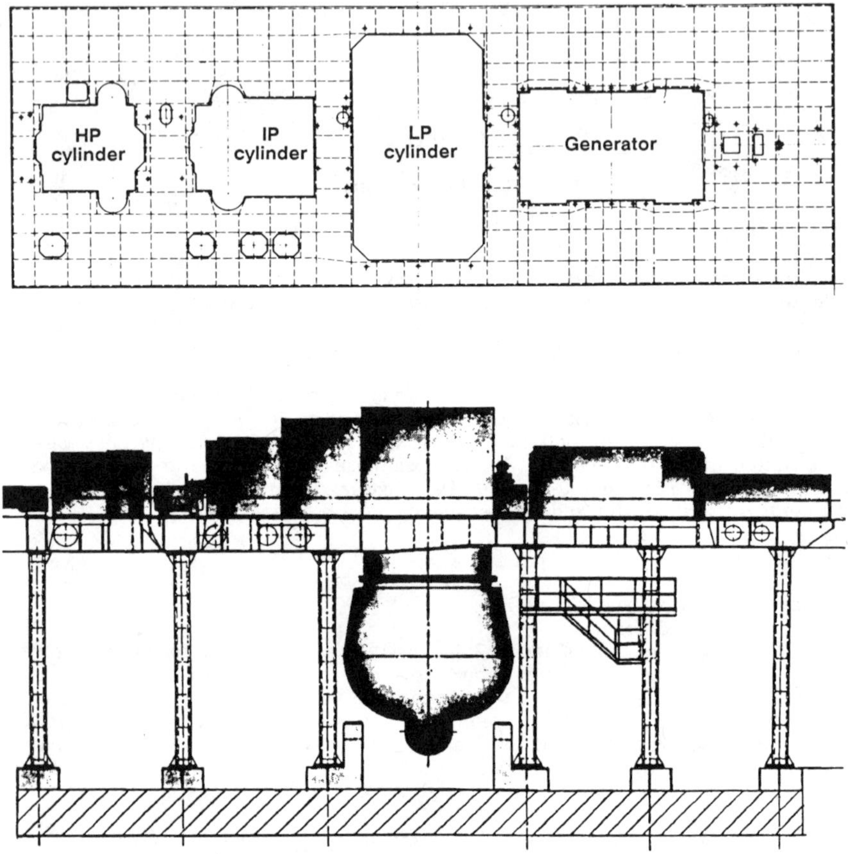

Courtesy of MAN GHH, West Germany

6627-27 2-86 KPS

Figure 25. Steel Foundation

Photo courtesy of MAN GHH, West Germany

Figure 26. Arrangement of Stiffeners in Steel Foundation Deck

Foundation Columns

The columns determine the fundamental frequency of the foundation and thus demand a certain cross section for the vertical spring action and a bending resistance for the transverse natural frequencies. They are also shaped as cellular structures to obtain the most favorable relationship between cross section and moment of inertia. The length of the columns (i.e., the height of the foundation) is determined by the local requirements. The lateral column dimensions are always smaller than those required for concrete foundations, which is the space-saving feature of steel foundations.

Erection Duration and Procedure

The field erection of a steel foundation takes only a few weeks. In the case of a 500 MW-unit, it may amount to 8 to 10 weeks. The erection of the steel foundation is a procedure that can be fitted almost anywhere into the overall construction schedule of a power station. Steel erection can, for instance, run simultaneously with part of the erection of the condenser.

Prefabricated parts of the steel foundation are shipped to the site and then assembled. The size of the individual prefabricated components depends on the mode of transportation. In general, the turbine room will be erected first, making the turbine room crane available for the erection of the steel foundation. However, if the turbine room crane is not available, erection of the steel foundation can be carried out with a mobile crane. The welding of the different parts is done under a sequence that ensures a structure within predetermined tolerances. After welding, all bearing plates are generally checked for levelness, squareness, and proper elevation. Distortions of the prepared bearing plates can be corrected by grinding to achieve a tolerance as small as 1:10,000.

Conclusions

The steel foundation has had a sufficiently long history to prove that it is an acceptable alternative for supporting large turbine-generators. The Committee is aware of only one firm that designs and fabricates steel foundations; this firm is located in Germany. The manufacturer claims many advantages for the steel concept over the concrete foundation.

Two clear advantages of steel foundations are apparent. The steel concept provides the greatest amount of free space below the deck for locating equipment. The steel concept also allows for major realignment of the turbine deck by shimming at the column baseplates, a procedure which could be necessary in the event there is significant foundation settlement. The overall advantages of concrete versus steel turbine-generator foundations must be evaluated on a case-by-case basis.

Chapter 9

Spring-Mounted Machine Foundations

Introduction

The use of spring-supported turbine-generator sets seems to be more common, particularly in Europe, than when they were first used in Germany in the 1960s. They have since been used in many different countries in Europe and in Canada and Australia. There are no spring-supported units in the United States. Two different types of spring units are used to support turbine-generator sets: helical-spring units and Belleville washer units. These two types are explained in the following sections.

Helical-Spring Foundations

The foundation support units are comprised of groups of cylindrical coil springs with circular cross sections (Figure 27). Such helical-spring units are available in various standardized design configurations and sizes with a load-

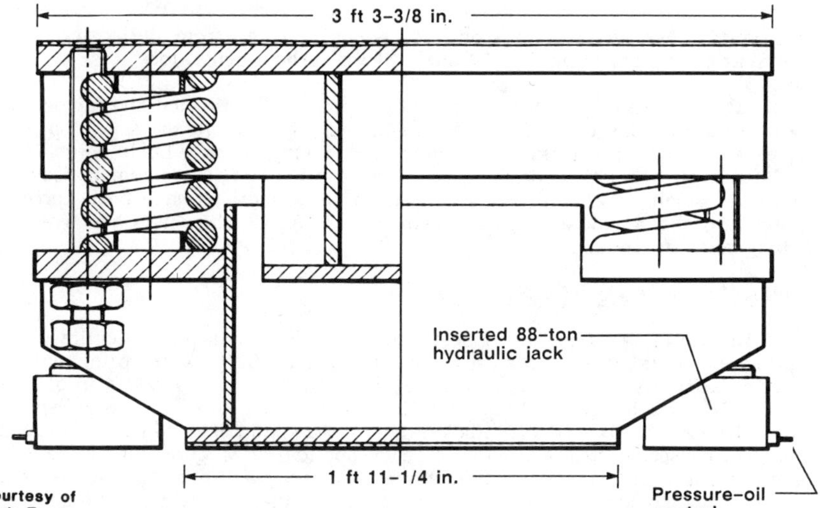

Courtesy of
Gerb, Essen

6627-29 2-86 KPS

Figure 27. Sectional Drawing of a Typical Helical-Spring Unit with Built-in Viscodamper at Center

bearing capability that ranges from 55 to 140 tons (500 to 1250 kN) each. Helical springs possess the advantage that they have precisely defined stiffness constants in both the vertical and horizontal planes. Lateral flexibility is important in eliminating any problems with mechanical and thermal forces in the horizontal plane. The helical-spring units are generally characterized by a stiffness ratio of roughly one, i.e., their vertical and horizontal spring constants lie in the same range. The fundamental natural frequencies of helical-spring foundations are generally about 150 r/min both vertically and horizontally. The static deflections of normally loaded helical-spring units are typically somewhat in excess of 1 inch (25 mm). Such large deflections are highly advantageous in minimizing the effect of substructure subsidence on the foundation deck elevation, and thus on the machine alignment.

The foundation designer determines the number of helical-spring units of different load-carrying capability that are required to support both the reinforced concrete foundation deck and the turbine-generator on the basis of the calculated portion of this combined load that has to be borne by each of the substructure columns.

If an expansion joint is provided between the low-pressure turbine exhaust and the condenser neck, the same spring units are also designed to withstand the resulting vacuum forces.

Each helical-spring unit is prestressed and locked in the field to simulate its predicted static deflection when loaded by the foundation deck and machine plus an extra amount, typically 20% to 30% of this computed deflection, to prevent further deflection during erection of the turbine-generator as a result of any abnormal distribution of weights on the deck (Figure 28). This prestressing of the units is designed to prevent the helical springs from deflecting during installation and affecting the alignment of the turbine-generator.

After the turbine-generator has been completely installed and aligned, but before the machine rotor couplings are bolted together, all the spring units are released of any prestress by opening the locking bolts. Each unit is then deflected strictly in accordance with the portion of the combined deck and machine weight it actually supports. If these deflections have been correctly predicted, the machine alignment will not be appreciably disturbed when the spring units are unlocked.

Belleville Washer Foundations

An alternative type of support utilizes conically shaped Belleville washer spring elements of high load-bearing capability. Such plate-spring units of varying stiffness are available in sizes that can each be loaded up to 550 tons (5000 kN). Each unit contains a certain number of Belleville washer springs, typically four or six that are stacked together in pairs back to back, to provide the desired spring characteristic for each application (Figure 29). In addition, each unit has its own self-contained hydraulic jacking device. The fundamental natural frequencies of plate-spring foundations in the vertical plane lie generally in the region of 200 to 350 r/min. The static deflections of normally loaded plate-spring units are usually less than 0.5 inch (13 mm).

Two separate types of plate-spring units are employed to support the weight of the reinforced concrete foundation deck and the weight of the turbine-

Figure 28. Helical-Spring Unit Being Prestressed Prior To Installation in Foundation

generator itself. If an expansion joint is installed between the low-pressure turbine exhaust and the neck of the condenser, which imposes vertical vacuum forces on the foundation, a third type of plate-spring unit is used to withstand this additional loading (Figure 30). The foundation designer selects the number of supporting units of various load-carrying capabilities of each type that are required on the top of each of the substructure columns to support the distributed foundation deck and machine weights and, if necessary, to withstand the calculated vacuum loads in the vertical plane. In addition, different vertical spring constants are prescribed for each type of spring unit at the individual locations, to minimize the deformation of the foundation slab. This involves providing stiffer plate-spring units at the center of the foundation in the region under the low-pressure turbine than towards the ends of the foundation beneath the high-pressure turbine and the generator and exciter.

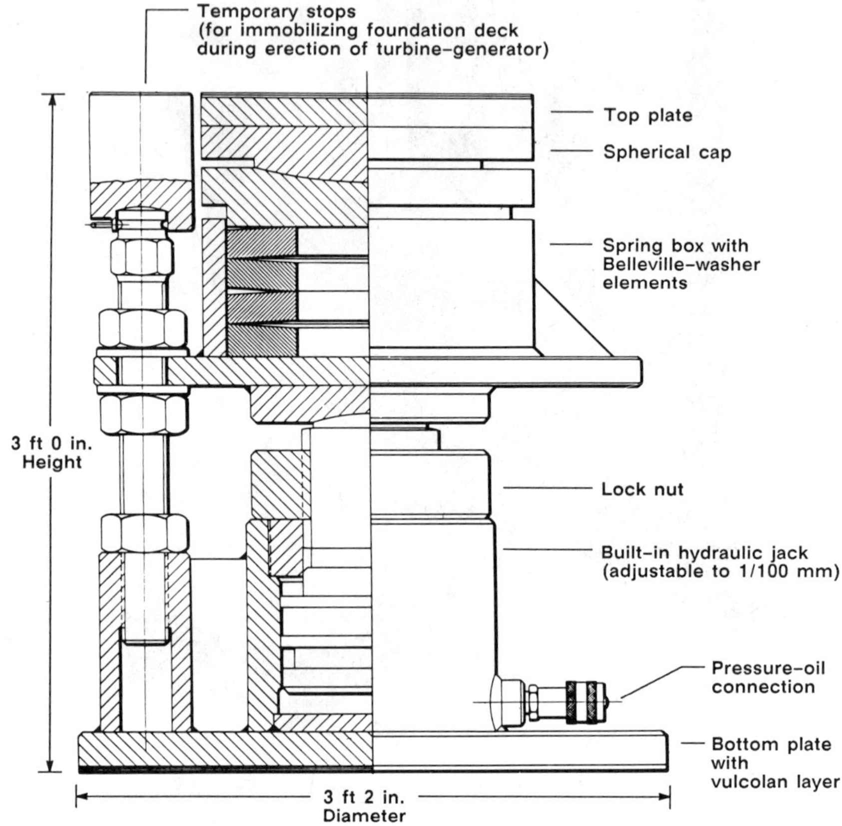

Courtesy of
Trombik & Stadelmann, Zurich

6627-31 2-86 KPS

Figure 29. Sectional Drawing of a 550-Ton Plate-Spring Unit with Associated Temporary Stop

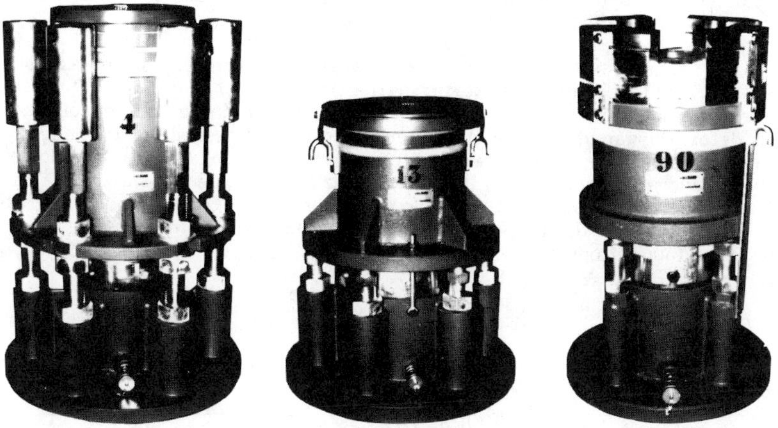

Foundation deck | Turbine–generator equipment | Condenser vacuum forces

6627–32 2–86 KPS

Figure 30. Three Types of Plate-Spring Units

All the plate-spring units are placed on top of the haunches of the foundation substructure columns and covered by steel plates that are integrated into the foundation deck formwork. The units that are provided to support the foundation deck and the turbine-generator are not prestressed. Only the units that are to withstand any vertical vacuum loads are installed with a prestress that is generally selected to correspond to the deflection that is anticipated to occur when the vacuum forces amount to 80% of their maximum value.

When plate-spring foundations are designed to withstand moderately severe seismic forces, their decks are connected to the surrounding turbine house floor by horizontal tie bars that are fitted with a special plate-spring unit at the end of each bar (Figure 31). The spring characteristic of the combination of each tension bar and its Belleville washer elements is selected to prevent the foundation deck from impacting against the operating floor as a result of the horizontal acceleration associated with a design-basis earthquake. In addition, the standard plate-spring units beneath the foundation deck are mounted on pairs of Teflon-coated, stainless-steel plates to reduce the lateral displacement of the deck in response to horizontal seismic acceleration of the supporting substructure (36). A plate-spring foundation design concept is also available to withstand very severe seismic forces. This concept locates the plate-spring units on a low-friction Teflon sliding surface at the base of each foundation column in the basement and also utilizes horizontally arranged tie bars with Belleville spring elements between the deck and turbine-house floor.

Experience with Spring-Supported Foundations

The operating experience with both helical-spring and plate-spring foundations has been excellent. No failures of any spring elements supporting turbine-generators have ever occurred in service. Several very serious incidents (for example, loss of turbine last-stage blades, generator terminal short circuits, and out-of-phase synchronization) were successfully withstood by large machines on spring-supported foundations.

Many measurements have been taken on various spring-supported foundations to ascertain the static deflections of their decks and turbine-generators under different operating conditions. Spring-mounted foundation decks deform dynamically in a characteristic way that resembles the undulating ripple patterns of a "flying carpet." This changing location of the lowest-frequency vibration nodes along the reinforced concrete decks is attributable to the fact that they simply rest on their spring units and thus are not integrated with the columns of the supporting structures. Additional information on these spring-type foundations can be found in References 13, 16, 22, 28, and 29.

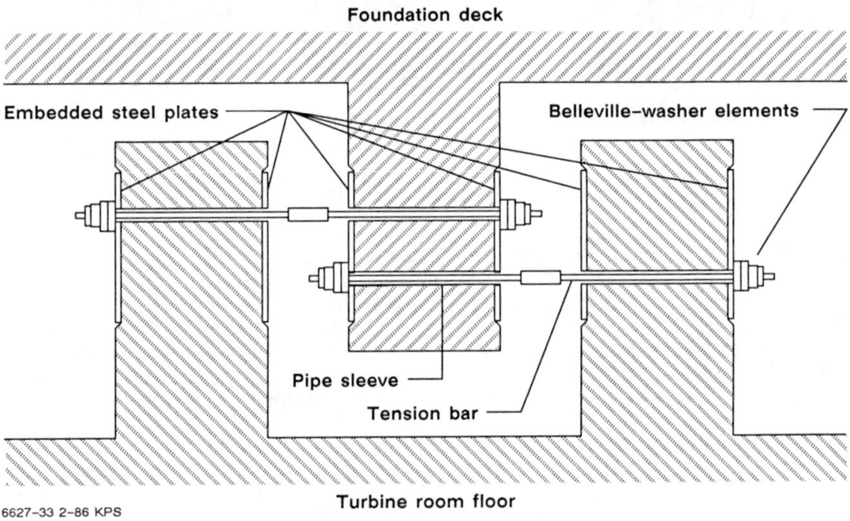

Figure 31. Typical Horizontal Spring Support of a Plate-Spring Foundation Designed To Withstand Severe Seismic Forces

Chapter 10

References

1. Aboul-Ella, F., and Novak, M., "Dynamic Analysis of Turbine-Generator Foundations," presented at the October 29-November 3, 1978, Fall Convention, ACI, held at Houston, Texas.
2. ACI Committee 436, "Suggested Design Procedures for Combined Footings and Mats," Journal of the American Concrete Institute, Vol. 63, No. 10, Oct., 1966, pp. 1041-1057.
3. Almuti, A. M., "Large Flexible Turbine Foundations," Methods of Structural Analysis, ASCE, New York, New York, 1976, pp. 707-719.
4. American Concrete Institute, "Building Code Requirements for Reinforced Concrete (ACI 318-83)," ACI, Detroit, Michigan, 1983.
5. American Concrete Institute, "Code Requirements for Nuclear Safety Related Concrete Structures (ACI 349-80)," ACI, Detroit, Michigan, 1980, Appendix B.
6. American Concrete Institute, "Testing Hardened Concrete: Non-Destructive Methods," ACI Monograph No. 9, ACI, Detroit, Michigan, 1976.
7. Aneja, I., "Dynamic Response of Systems-Turbine Generators on Various Foundations," Proceedings of the American Power Conference, Vol. 37, 1975, pp. 528-540.
8. ANSYS computer program, "Engineering Analysis System," Swanson, J. A., Swanson Analysis Systems, Inc., Houston, Pennsylvania.
9. Archer, "Consistent Mass Matrix for Distributed Mass Systems," Journal of Structural Division, American Society of Civil Engineers, Vol. 89, No. ST4, August, 1963, pp. 161-178.
10. Arya, S., and Drweyer, R., "Mathematical Modeling and Computer Simulation of Elevated Foundations Supporting Vibrating Machinery," Transactions of IMACS, Vol. xix, No. 4, December, 1977.
11. Arya, S., O'Neill, M., and Pincus, G., Design of Structures and Foundations for Vibrating Machines, Gulf Publishing Co., Houston, Texas, 1979.
12. Benuska, K. L., Chen C., and Miller, R. K., "Analytical and Experimental Investigations of the Martins Creek Low Tuned Concrete Turbine Pedestal," presented at the February 20-21, 1975, Pennsylvania Electric Association, Structures and Hydraulics Committee, Winter Meeting, held at Bethlehem, Pennsylvania.
13. Danisch, R., and Labes, M., "Aseismic Design of Turbine Houses for Nuclear Power Plants," Nuclear Engineering and Design, Vol. 38, No. 3, 1976, pp. 495-501.
14. Hadjian, A. H., "Design Criteria for Turbine-Generator Pedestals," Journal of the Power Division, American Society of Civil Engineers, Vol. 96, No. PO1, January, 1970, pp. 1-22.
15. Hadjian, A. H., "Design of Pedestals for Machine Unbalance," presented at the September 10-14, 1972, ASME-IEEE Joint Power Generation Conference, held at Boston, Massachusetts.

16. Hueffmann, G., "Spring Mounting of Turboset Foundations," Klein, G., "Tuning of Foundations for Steam Turbine-Generators and Consequences for Power-Station Operations," and Trombik, G., "Spring-Mounted Foundations for Turbo-Generator Sets," papers contributed to October 29-31, 1973, 44th Euromech Colloquium: Dynamics of Machine Foundations, Bucharest, Romania.
17. International Conference of Building Officials, "Uniform Building Code," 1985 Ed., ICBO, Whittier, California, 1985, pp. 114-123.
18. ISO Technical Committee 108, "Balance Quality of Rotating Rigid Bodies (ISO-1940)," International Organization for Standardization, Geneva, Switzerland, 1973.
19. ISO Technical Committee 108, "Mechanical Vibration of Large Rotating Machines with Speed Range from 10 to 200 r/s - Measurement and Evaluation of Vibration Severity In Situ (ISO 3945)," International Organization for Standardization, Geneva, Switzerland, 1985.
20. ISO Technical Committee 108, "Mechanical Vibration of Machines with Operating Speeds from 10 to 200 rev/s - Basis for Specifying Evaluation Standards (ISO 2372)," International Organization for Standardization, Geneva, Switzerland, 1974 (Amended 1983).
21. Hurty, W. C., and Rubinstein, M. F., Dynamics of Structures, Prentice-Hall, Englewood Cliffs, New Jersey, 1964.
22. Kasten, H. L., and Kirkland, W. D., "Spring-Mounted Turbine Generator Spins Quietly, Efficiently," Electric Light and Power, E/G Edition, Nov., 1970, pp. 38-40.
23. Kleinlogel, A., Rahmenformeln, Springer Verlag, Berlin, Germany, 1949.
24. Leontovich, V., Frames and Arches, McGraw-Hill, 1959.
25. NASTRAN computer program, "NASA Structural Analysis," National Aeronautics and Space Administration, NASA SP-222, Goddard Space Flight Center, Greenbelt, Maryland.
26. Patterson, B., "Turbine-Generator Pedestal Design," Proceedings of the American Power Conference, Vol. 37, 1975, pp. 541-549.
27. Portland Cement Association, "Concrete Floors," 2nd Ed., PCA, Skokie, Ill.
28. Probst, P. H., "The Foundations of a 95 MVA Motor-Generator Set for Pulsed Duty," Seimens Review, Vol. 39, No. 1, 1972, pp. 28-31.
29. Probst, P. H., and Joyce, J. S., "The Development of Helical-Spring Foundations for Large Steam Turbine-Generators," Proceedings of the American Power Conference, Vol. 34, 1972, pp. 424-437.
30. SAP computer program, "Structural Analysis Program," Wilson, E. L. et al., University of California, Berkeley, California.
31. Shen, G. T., and Stone, N. E., "Natural Frequencies of a Turbine Foundation," Structural Design of Nuclear Plant Facilities, Vol. II, ASCE, New York, New York, 1975, pp. 302-330.
32. Srinivasulu, P., and Lakeshamanan, N., "Dynamic Response of Turbo-Generator Pedestals," presented at the April 24-28, 1978, ASCE Spring Convention and Exhibit, held at Pittsburgh, Pennsylvania (Preprint 3149).
33. STARDYNE computer program, "Structural Analysis System," Mechanics Research Inc., Los Angeles, California.
34. STRUDL computer program, "Structural Design Language," Massachusetts Institute of Technology, Department of Civil Engineering, Cambridge, Massachusetts.
35. Timoshenko, S. P., Young, D. H., and Weaver, W., Jr., Vibration Problems in Engineering, 4th ed., Wiley Publishing Co., New York, New York, 1974.

36. Trombik, G., "Spring-Mounted Foundations for Turbosets," paper presented at 1967 Fall Canadian Power Conference, held at Edmonton, Canada.
37. Volterra, E., and Zachmanaglou, E., Dynamics of Vibrations, Merrill Books, Inc., Columbus, Ohio, 1965.